The Cosmic Bridge

Close Encounters and Human Destiny

Craig R. Lang

With a foreword by Raymond E. Fowler, Author of *The Watchers* and *The Allagash Abductions*

Copyright Page

To Gwyn, who has put up with the many hours in which I have been "abducted" by my UFO studies.

Table of Contents

Foreword

By Raymond E. Fowler

Humankind from all ages has reported strange objects in the sky. Each generation has attempted to explain them within their own cultural framework. Earlier non-scientific cultures equated them with their particular religious belief systems. Meteors, aurora borealis, sundogs, moon dogs, eclipses and especially comets were supernatural in the eyes of the beholder. It is possible that what our generation call UFOs were also reported in centuries past but this is pure conjecture. UFO, according to the United States Air Force "relates to any airborne object which by performance, aerodynamic characteristics, or unusual features does not conform to any presently known aircraft or missile type, or which cannot be positively identified as a familiar object." [AFR 200-2]

UFOs first drew widespread attention in modern times when newspapers publicized the sighting of nine flat shiny objects by Kenneth Arnold, an experienced pilot, on June 24, 1947. He reported that they flew with an up-and-down motion like a saucer skipping over the water. The press dubbed them flying saucers or flying disks. Most reports could be explained as misinterpretation of natural phenomena or misidentification of common man-made objects but at times up to 20% remained unidentified after investigation by the Air Force.

Modern UFO sightings began during World War II and were first thought to be enemy secret weapons. Small disk and globe-shaped objects paced both allied and enemy aircraft. After the war, thousands of disk and cylindrical-shaped objects were reported over Scandinavian countries. In 1947, the first of many waves of UFO sightings took place all over the world. The then Army Air Force took steps to identify the strange objects.

On September 23, 1947, General Twining in a now declassified secret memo wrote that:

"The phenomenon reported is something real and not visionary or fictitious. There are objects probably approximating the shape of a disc, of such appreciable size, as to appear to be as large as man-made aircraft. Action, which must be considered evasive when sighted or contacted by friendly aircraft and radar, lend belief to the possibility, that some of the objects are controlled manually, automatically or remotely. The apparent common description of the objects is: circular or elliptical in shape, flat on bottom and domed on top."

The United States Air Force then expanded the investigation. Project Sign was initiated in 1948. Sign's project chief, Edward J. Ruppelt admitted that the project's estimate was that UFOs were interplanetary spaceships. However, this estimate was rejected; the public was told that all UFOs could be explained and that the investigation was over. However, a new secret project, entitled Project Grudge, continued the investigation. It surfaced publicly during a huge wave of UFO sightings in 1952 and was subsequently entitled Project Bluebook. Bluebook continued until it was terminated in 1969. The government now continues investigations without public participation.

Due to national security, the public found it hard to obtain information about UFOs from government sources. This resulted in a number of civilian investigation groups being formed which conducted their own investigations and reports to interested civilians. After following the subject for decades, the author of this book became a field investigator for one such group, an international organization called the Mutual UFO Network (MUFON). I am well acquainted with this organization and served it as Director of Investigations on the Board of Directors. In order for him to become an investigator, he had to pass a detailed exam based upon the MUFON Field Investigator's Manual. Lang is well educated. He has a Masters Degree in software engineering and is a certified hypnotherapist. Both of us as well as other UFO researchers have written articles and books to keep the public informed about UFOs.

During our investigations, we have seen a steady progression in how this strange phenomenon has exposed itself to human eyes over the years. First, there were objects sighted in the distance. Then there were close-up views of metallic machine-like objects. Later there were sightings of objects landing and leaving physical traces behind in their wake. Soon after, there were sighting reports of alien beings associated with the objects. Then and now, there are the UFO abductions of witnesses who are taken aboard, given physical examinations and sometimes messages and then returned to the point of their capture.

Lang documents that such close proximity to UFOs and their operators have opened those abducted to a number of alternate realities, which appear to be part of a conditioning process. It was and is the exponential exposure of the UFO phenomenon to humankind that instigated the author to write this book.

Lang believes that the phenomenon is preparing humankind for ultimate overt contact sometime in the future. Lang envisions their step-by-step revealing of themselves over the years as a Cosmic Bridge to an alien civilization's accepting us into an intergalactic society.

Lang's hypothesis is not just based upon the investigations and evaluations by others. It is also based solidly on his first hand investigations of the progressive types of UFO experiences – from distant sightings to intimate alien contact - that form the rationale for his hypothesis. Lang painstakingly mimics the phenomenon itself by guiding the reader step-by-step through the variety of UFO experiences that are currently in the process of building a Cosmic Bridge to human destiny.

Preface, a Personal Journey

Alien abduction - the very words suggest one of the greatest mysteries of our time. Who are these visitors from the sky? Why are they here? For me, this question has been the spark behind a lifelong journey.

For as long as I can remember, I have been interested in space, UFOs and the possibility of extraterrestrial life. I was probably too young to remember my first exposure to the idea of alien contact. Yet my first childhood memories involve an interest in space. The cartoons which I watched as a child, the books I read, even my banter with playmates, nearly all involved the idea of space travel and/or aliens.

I was always imagining what it would be like to travel to the stars, or to meet those who had traveled from them to Earth. Never for a moment did I doubt that star travel was possible. In fact, I always assumed that the presence of visitors from the stars was a reality. For me, this was simply an article of faith.

In the mid 1960's, my grandmother gave me a copy of the book, *The Interrupted Journey* by John C. Fuller. This was the story of the Hill abduction case and was my first exposure to the UFO encounter narrative. For me, it began what would become a life-long passion.

Not long after, I also read about the first serious efforts in the Search for Extraterrestrial Intelligence (SETI). This appealed immediately to my imagination. Thus, during the early 1970s, I became actively interested in astronomy and electronics.

Throughout the years, the UFO phenomenon has exhibited many peaks and dips. There were the flying saucer sighting waves of 1964 and 1968. In 1973 there was the "Year of the Humanoid" and the Pascagoula, Mississippi abduction of Charles Hickson and Calvin Parker. I closely followed the sighting waves of 1978/79, including the apparent collision of Deputy Val Johnson's car with a UFO in Minnesota during the summer of 1979. The late 1970s was also when I first saw the movie *Close Encounters of the Third*

Kind, by Steven Spielberg. Each of these further sharpened my lifelong interest in both space and alien contact.

My interest in electronics led me to a career in electrical engineering. In 1980, I graduated in engineering from the University of Wisconsin and then worked for several electronics companies. I moved into computer engineering in the 1980s and have been a software engineer to this day. This path led me to graduate school, resulting in a master's degree in software engineering at the University of Saint Thomas, in Minnesota.

In 1996, on completion of graduate school, I also became active in The Mutual UFO Network (MUFON) as a field investigator. At that time, I began to interview abductees and became even more deeply intrigued by the mystery of alien encounters. As a result, I decided to study hypnosis as a better way to help me work with abduction experiencers. In 1998, I became a certified hypnotherapist and began to do regression work in UFO abduction cases. My journey of exploration into what may be the biggest mystery of modern times had taken an amazing new turn.

Working with abduction experiencers has convinced me of the reality of their encounters. My work with hypnosis clients, studies of the literature and field investigation of sighting/encounter cases have continued to reaffirm the certainty of that reality. Yet I also see such encounters as a mystery that deepens at every turn. It has very much enhanced the sense of wonder that I experience when dealing with this topic. Yet I believe that the number of dead ends along the trail and the many logical paradoxes presented by the UFO phenomenon have also left me with a very discerning eye.

Whatever the true nature of the phenomenon, I have found that the encounters described by others seem to have a solidly objective aspect to them, a powerful quality that has led me to accept their reality for the experiencer. Abduction accounts frequently tend to correlate with physical evidence or other corroborating details. And they are nearly always recounted with the emotion that tells me that, for the experiencer, they are deeply real.

During the mid to late 1990s, as I also began to study spiritual and metaphysical topics, I came to the realization that one purpose for my study of the UFO phenomenon has been to forge a bridge to a new understanding of extraterrestrial contact. My primary mission, and the core purpose of this book, is to help bring the world of humanity and that of the phenomenon, whatever it may be, into a positive relationship. No doubt, I am simply one of many who are part of this endeavor. However, it is my firm belief that, for better or for worse, contact is beginning to expand and deepen.

We live in a time in which human and visitor consciousness are beginning to approach each other. I suggest that this process has already begun. Today, as a species we are beginning to build and to cross The Cosmic Bridge to the other side of the sky.

Acknowledgements and Dedication

This book is the result of many years of evolution. It has taken shape from my own research, thoughts and articles, as well as the input and assistance of many others. Many people were involved at each stage of developing this book.

First and foremost, I would like to thank the many sighting witnesses and close encounter experiencers whom I have had the pleasure of meeting. I thank them for offering me their stories. They include those whom, in this book, I have given pseudonyms such as "Amy," "Jenny," "Janice," "Carolyn," "Evelyn" and many others. Without their help, this work would not have been possible.

I would also like to thank Kevin Hogan and C. Devin Hastings, my instructors at the Minnesota Institute of Hypnosis and Hypnotherapy. It was with their help that I began my hypnotherapy career and thus, my in-depth work with close encounter experiencers. Their tutelage has enabled me to accomplish that which I have done to date.

I wish to thank everyone at the Loft Literary Center in Minneapolis, Minnesota, for their instruction on various aspects of writing, editing, publishing, etc. I would also like to thank the many people whose comments were invaluable in the writing and editing of this book. Thanks go to Michael Lindemann, who commented on early drafts of this book and whose discussions of UFOs and contact provided me with both guidance and inspiration. I also thank Joan Hause, professional editor, who thoroughly edited this book for both form and content. Thank you also to Dean and Margaret DeHarpPorte, Richard Moss, Bill McNeff, Joe Dundovic and Solara An, all of the Minnesota chapter of the Mutual UFO Network. The feedback from each has been extremely helpful in the development of this book.

I further thank the many field investigators and participants of the Mutual UFO Network who have provided an invaluable service both to UFO research and to the many experiencers of the close encounter phenomenon.

Many thanks also go to my family - to my mother, Jeanne Lang, who awakened in me the imagination and inspiration to explore the unknown, and to my father, Raymond Lang, who instilled in me the drive and the common sense needed to complete this project.

Special thanks go to my late grandmother, Evelyn Herrling. Her giving me the book, *The Interrupted Journey*, which compellingly describes the abduction of Barney and Betty Hill, set me on a path that has altered my life to this day.

Finally, I would like to thank my wife Gwyn for her understanding and patience during those times in which I have been "abducted" by my UFO studies. Her support has truly made this book possible.

To Gwyn I dedicate *The Cosmic Bridge*.

Introduction - The Other Side of the Sky

I love the sky, deep blue by day, dark and infinite by night. To me it is a source of imagination and fascination. Yet for some, the sky holds a deeper meaning, sometimes wondrous and other times sinister. It is the realm of the close encounter phenomenon and the domain of the visitors.[1]

The UFO phenomenon is one of the longest lasting, deepest and most perplexing mysteries in human history. What are these strange things in the sky? How and why are they interacting with us? There is intelligence behind them, but who or what is it?

For many in society, encounters with this strange influence pervade life. For them and thus for us, the phenomenon of UFOs and encounters with visitors from somewhere else is a reality.

A recent poll of unusual personal experiences estimates that nearly one person in a hundred has experienced the close encounter phenomenon in some way. For some, such an experience might be an interesting mystery to explore. For others it may be a deeply disturbing unknown. It may exist as a long-standing thread of unexplainable events coloring one's relationships, altering one's views of the world and even affecting one's very health. For still others, encounters may be wondrous, a source of contact with the amazing. In one way or another, each experiencer becomes part of a process, a relationship with the extraordinary.

The UFO and Close Encounter phenomenon has been an underlying passion of my life for many years. It has taken me through many mental and spiritual adventures. As a UFO investigator with the Mutual UFO Network,[1] I have interviewed scores of UFO witnesses and close encounter experiencers. People have described events that span the full range - from seeing lights in the sky to experiencing personal encounters with the unknown.

---

[1] In his book, *Communion* [Avon, 1988], Whitley Strieber refers to the intelligence associated with the UFO phenomenon as "the visitors." Regardless their nature and origin, we will refer to them by this term throughout this book.

As I began my work as a hypnotherapist and began to study the phenomenon more deeply, I noted how the encounters people described took on very personal aspect. They were deeply real and complex. What at first seemed to be a simple, scattered phenomenon I have come to see as one of tremendous depth, permeating our society. In it, I see deep implications for the personal reality of all of us.

What is this close encounter phenomenon? Are the visitors physical beings from somewhere else, or is there another explanation for their origin? Do they mean us good or ill? There is a broad spectrum of views, some seeing them as good while others see them as evil. Their intentions appear to be complex, indeed.

In this book, I suggest a comprehensive model of the phenomenon and its interactions with humanity. I propose that the visitors are here for a very distinct reason, the preservation of the cosmic community.

I propose that the ultimate objective of the visitors is to build a long-term bond with our human race - a race that is steadily advancing in numbers and technology, preparing to travel to the stars.

Many writers on extraterrestrial contact have described humanity as a primitive people, yet also promising. Artists, healers, priests, warriors and scientists; we are all these things and so much more. I believe that we can be the bearers of a new creative light to the universe, or we can bring destruction. I propose that it is the visitors' intent to ensure that we are the former.

I suggest that the purpose for the visitors' presence is to nurture a human race more capable of becoming citizens of a universal community. I further suggest that their ultimate goal may be to see human consciousness somehow converge with that of the cosmic community - the building of The Cosmic Bridge.

This book presents ideas of our relationship with the visitors. What is it, what has it been and what is it becoming? It is a synthesis of what I believe experiencers and the UFO literature have told me. It is my hypothesis about what is going on between the visitors and humanity.

What is a hypothesis? It is not a statement of belief, but rather, a model of what might be the truth. A hypothesis must provide predictions. What must we see for the model to be correct? We can verify it by experiment, observation or other means of evaluating how our universe and its occupants actually behave. In this way, we can confirm or refute our original premise.

In the first chapter, I discuss the basics of UFO studies and describe some important and spectacular UFO sightings. We begin to see the UFO interaction as a relationship. In the second chapter, I discuss UFO abduction and describe a number of the key abduction experiences in the lives of experiencers. In the third chapter, I expand the exploration of close encounters, to illustrate the high strangeness that the phenomenon exhibits. In Chapter 4, I then discuss the impact the phenomenon has on the lives of experiencers and how the phenomenon is, in fact, a lifelong relationship.

Chapter 5 describes the human-visitor relationship in more detail, focusing on the growth and learning that can occur, both because of the phenomenon and as a way of coping with the inescapable influence of the visitors. Chapter 6 then focuses on the reproductive and genetic aspects of the phenomenon, laying the groundwork for our discussion of a possible alien agenda.

The latter half of the book discusses the overall human-visitor dynamic, moving from the individual level to the societal level. In Chapter 7, we review some conventional ideas regarding contact and examine the potential dangers of direct contact with ET visitors. In chapter 8, we delve into the heart of the topic, discussing possible goals of the visitors and their potential strategies for accomplishing them.

Chapter 9 expands our discussion of possible directions of the phenomenon and the promises and consequences of our relationship with the visitors. We conclude with a summary of what I believe may be the evolving human-visitor relationship. As our civilization matures, I envision a deepening link between humanity and our celestial neighbors.

This book makes some significant assumptions that the reader must decide whether to accept. The first is that the UFO phenomenon is real and that accounts related by close encounter experiencers are valid. The second assumption is that parapsychological effects (telepathy, clairvoyance, precognition, etc.) and events such as out of body experiences (OBEs) and spontaneous human invisibility are bona-fide phenomena. Many of these phenomena far transcend the UFO/close encounter phenomenon and are, themselves, significant mysteries.

In many events referred to in this book, I have obscured case details to preserve the anonymity of the witnesses and/or experiencers. In several instances, I have aggregated more than one case, event or person together into a representative account. However, the events themselves are accurate and fully illustrate the points within The Cosmic Bridge, portraying the building of the human-visitor relationship.

The visitors can be our teachers and guides or they can be our nemesis. To some degree, the future of our relationship with them is up to us. Whatever that future, I suggest that they may be the bridge to the destiny of humanity: the Cosmic Bridge to other side of the sky.

Chapter 1 – The UFO Phenomenon in Perspective

Imagine that you are alone in your car, on the highway in the dark of night. A mile or so ahead, you can see a light. It is not familiar looking. Maybe it's just an airplane. Yet as you drive further, you see that the light has descended toward the road. As it gets closer, you see it more clearly. You realize that it is not an airplane and not a car. Indeed, it is nothing you have ever seen before...

A UFO encounter might begin as a sighting during a drive, but then result in a missing hour or two. Or perhaps an encounter might take the form of a strange surreal nightmare, a vague memory of uninvited visitors during the night. There are many possible encounter scenarios and each brings us a glimpse into the extraordinary, a footstep into the unexplained.

How many people have had such an experience? Estimates suggest that five to ten percent of the population have witnessed something unexplained in the sky and nearly one percent of us may actually experience close encounters.[ii] Thus, for many, the UFO phenomenon is undeniably real. It is therefore vital that we more clearly understand our visitors from the sky.

What are the visitors doing? Why are they here? What is our relationship with them? These are perhaps the defining questions of what may be the greatest mystery of our time. This book is about that relationship - the bridge between humans and visitors from beyond Earth. In these pages, we consider this, first for the individual experiencer, then for humanity as a whole. For this relationship forms the cornerstone of the Cosmic Bridge between their realm and ours.

As we will see throughout this book, the UFO phenomenon can involve events with a vast range of meaningfulness, from the simple sighting of a light in the sky to a full-blown UFO abduction. Yet, regardless of its nature, nearly every encounter begins with the sighting of an object or entity. Thus, to understand our relationship with the visitors, it is vital that we first see UFO sightings in perspective. We therefore begin with a consideration of the sighting experience.

UFO 101

What is a UFO sighting? In the language of today, the term UFO conjures up visions of flying saucers, extraterrestrials and other extraordinary phenomena. But what does a UFO event actually mean?

First and foremost, a UFO - or Unidentified Flying Object - is anything unidentified in (or from) the sky. UFO encounters can range from a distant sighting, perhaps a far-away object that looks almost explainable, to a life-changing event such as contact or abduction [See Appendix B - The Vallee Classification System of UFO Sighting Types and Kinds].

Most importantly, we can think of the UFO encounter as a relationship between the witness and the phenomenon. In the simplest case, in which a distant light is seen in the sky, such a relationship is simple and one-way. There is no apparent interaction with the phenomenon itself. The phenomenon is a temporary curiosity, witnessed for a moment.

In closer encounters, the relationship deepens. The interaction between the witness and the UFO becomes more meaningful and the strangeness - the degree to which the encounter departs from the witness's familiar reality and understanding - becomes greater. When a Close Encounter of the fourth kind (CE4) occurs, the witness becomes an experiencer, a full participant in the reality of the phenomenon. Such experiences include UFO abduction and we will see in later chapters how these form one of the key elements of alien contact.

Distant Encounters: A Classic Sighting in St. Paul

Distant encounters are perhaps the most common type of UFO sighting. We indicated earlier that up to one person in ten has sighted some form of distant, unexplained object. However, one example from my own case files best illustrates some of the interesting aspects of such cases.

In this sighting, the witness was having a late lunch at a restaurant in St. Paul, Minnesota. This particular witness indicated that he frequently looks up at the sky and at this moment, found himself briefly looking out of the window near his table at the puffy clouds of a beautiful fall day. As he looked up, an object caught his attention. He initially thought it was an airplane. Yet it appeared to be motionless for what seemed to be a long time.

The object further attracted his curiosity and as he watched, it slowly began to descend at a 45-degree angle to the right. When it had descended nearly to the city skyline, it suddenly and abruptly shot upward. The witness followed the object with his eyes until it reached (and apparently passed behind) a small cloud. He estimated the vertical motion as well more than twice as fast as the descent and, as far as he could tell, the shift in direction was instantaneous. It was a maneuver impossible for a human-made craft.

While the description of the object and its behavior suggests that it is a genuine unknown, in this event the level of strangeness is relatively low. The witness did not interact with the object. It is a classic distant encounter with little overall impact on the witness's life-path.

This event was but one of several UFO encounters he had had. Yet even though this event had relatively low overall significance, it has continued to intrigue him. It illustrates how the phenomenon can attract and hold our attention and yet remain unexplained. This and so many similar cases add to the abundance of evidence that extraordinary things are common in our skies.

Close Encounters of the First Kind: The Medicine Lake Triangle

In a close encounter of the first kind (CE1), the level of meaningfulness increases, although the witness still has only limited interaction with the phenomenon. While the witness remains separate from the UFO, he/she now observes the object at close range.

Examples of close encounters of the first kind commonly include cases in which the witness observes a UFO flying or hovering nearby, perhaps only a block or two away, or in some cases even closer. One classic example of a close UFO sighting comes from the case files[iii] of the Minnesota chapter of the Mutual UFO Network, or Minnesota MUFON. In this event, a witness was driving along a lakeshore road in a rural/suburban region near Minneapolis. Looking up, she was surprised to observe a pair of bright lights ahead in the sky to her left, over the lake. They seemed to be out of place.

As she watched, it became apparent to her that the lights were not too far above treetop level. She continued to drive up the road for approximately one-half block, observing the lights through the trees the best she could. She then realized that the lights were considerably closer than she had thought.

After driving a few hundred feet further, she reached a point where a gap in the trees allowed her an unobstructed view. However, at this point the lights were no longer visible. In their place was a large dark triangular object.[2]

As she watched, the object moved slowly, just above the treetops. It was nearly the size of a house and was black to charcoal-gray colored with lights at the vertices. The object passed overhead, remaining at the same level, just above the treetops. As it did, she felt an intense mixture of awe and fear. It seemed to her as if the object (and/or those controlling it) were aware of her presence.

---

[2] Note: The black triangle is a commonly seen UFO.
See http://www.ufoevidence.org/topics/triangles.htm for description of this type of object.

She pulled her car ahead and turned into a driveway leading toward the waterfront. She finally got out and observed the object from shore as it slowly moved out over the lake. She watched it move away, eventually losing sight of it in the tree line along the distant lakeshore.

Just beyond the trees, only a few blocks away, a major highway passes near the lake. At this time of the evening, the highway was quite busy. The object, moving slowly at such a low level and approaching the highway, should have been clearly visible. However, as far as is known to investigators, this witness was the only person to see it - a selective viewability that seems to add yet another layer of mystery to the event.

In this case - a close encounter of the first kind - the witness observes much more detail than in a distant encounter. The level of strangeness becomes correspondingly greater. There is still little overt interaction between the UFO and the witness. Yet the witness's fear and simultaneous fascination with the object, plus her impression that the object was aware of her, suggest that here too, there may have been a subtle two-way interchange.

In addition, the object was visible to this witness but apparently not to others. Thus, more than a simple sighting relationship appears to have been present. It suggests that the nature of the event was more involved and our understanding of it was less than we thought.

Other Selective Viewability Cases

The case files of many UFO investigators are full of similar accounts of selective viewability. In such cases, the witness is able to see an object, which is located in such a way that many others present should also see it, but do not.

- During a power failure in South Minneapolis, on a warm autumn evening in 1995, a witness looked out of a second story window to see many people on the street below enjoying a warm, but suddenly dark, evening. Above, apparently unnoticed by them, a large UFO was hovering over the buildings only a few blocks away. The object was located such that it should have been obvious to those standing outside. Yet no one on the street appeared to be aware of it.

- On an early summer evening in 1996 a mother and her adult daughter were driving along a major highway in the northern suburbs of Minneapolis, Minnesota. The mother was driving as the daughter, in the passenger seat, looked up into the distance to see what initially appeared to be a bright light. She looked away briefly to point it out to her mother. When she looked back, instead of the light, there was a large dark triangle passing over the heavily traveled highway. They pulled the car off the road and watched as the huge, silent object passed slowly overhead at treetop level in full view of the motorists, who seemed oblivious to its presence.

Such case reports fill my files as well as those of other researchers. They lead me to ask whether in each case the witnesses possessed some unique ability to see the object, an ability not present among other bystanders. With the potential of this perception ability, the relationship of a simple sighting begins to turn into the increasingly meaningful relationship known as a close encounter.

Close Encounters of the Third Kind: The 1969 Race Track CE3

During a close encounter of the third kind (CE3), in addition to the UFO itself, the witness also observes entities associated with the object. Although the witness remains separate from the UFO, we continue to note the increasing level of interaction between them. One example of such an encounter from my own case files seems both very simple and very typical. In this case, the witness (whom I will call "Janice") briefly observed several entities within a UFO.

As a child, during the late 1960s, Janice and her father had attended an auto race. She was sitting on the top row of the bleachers, looking down onto the racetrack. At that moment, she got the sensation of someone or something watching her. Looking to her right, she observed an object hovering above the treetops less than 1000 feet away. She turned to point the object out to a person next to her, but when she looked back, it was gone.

Janice remembered the object as having a diameter of about 30 feet. Within it, she could see three or four entities in the window. In her fleeting glimpse, she was unable to note many details. However, she clearly felt the impression that they were observing her.

Close encounters of the third kind seem to strike a delicate balance in which the witness observes entities but does not directly interact with them. Yet we note Janice's sense of being watched, suggesting that the phenomenon was aware of her. Thus, the relationship appears to have been two-way, with object and witness observing each other.

Again, we note the witness' increasing relationship with the phenomenon. Our simple UFO sighting is becoming an encounter experience.

Close Encounters of the Fourth Kind, From Sighting to Experience

So far, in this chapter we have examined increasingly meaningful UFO sightings. With each, the witness has become closer to the UFO and the level of strangeness has increased. We have increasingly seen it as a relationship between the witness and the phenomenon. In a sighting or encounter of the first through third kinds, this interaction is distant and usually subtle. However, in a close encounter of the fourth kind (CE4), a UFO abduction, the relationship becomes up close and personal. The witness has now become an experiencer.

The Laundry Room CE4

One classic encounter report from my own case files is that of an experiencer I will refer to as "Evelyn." I first met Evelyn in 1997, when she reported to MUFON accounts of her many strange events. Shortly thereafter, we began our extensive interviews with her and her family and her long history of encounters emerged. One of the first she recounted was an experience that had left her with a deep unsettling sense of mystery.

In the dark of a late November night, there are few lights and little illumination in the rural northern Wisconsin landscape. It was well after midnight and everyone was, or should have been, sound asleep. In one farmhouse, only a few hundred yards from the main highway, the lights were off. Only a distant yard light shone on a farmyard ready for the first winter snow.

Evelyn awoke at about 3:30 AM. An eerie sense filled her as she sat up in bed. Nothing seemed to be amiss, yet something suggested that she needed to get up. She decided to go to the bathroom, stood up and stepped into the hall.

The bathroom was located near the north side of the house adjacent to their laundry room. From that point, a window looked out to the back of the house, affording a view over the back yard and the fields to the north. The field was dark, but beyond a grove of trees on the far side of the field, Evelyn could see a distant light. Though she had lived in this house for many years, the light was not something she had ever seen. It seemed to silhouette the trees with increasing brilliance. As she watched, she felt ever more uneasy about what she was observing.

The light seemed to be rising from behind the trees. She then realized that there were multiple lights, part of a large object emerging above the tree line. It was fully visible, a formation the size of a house and in the shape of a flying saucer. As she watched, it began to move toward her home.

As the object approached, she became increasingly frightened. She started to lock the doors and windows of the house, beginning with the laundry room, then the back door and then the windows in adjoining rooms. Finally, she ran to the front to lock the kitchen door, but it was too late. "They" were already there...

As she reached the kitchen door, she realized it was starting to open, which meant to her that the entities were beginning to enter the house. She braced herself against the door to hold it shut, but felt a force pushing it open. A moment later, she found herself standing behind the fully open door.

With the door open, she observed lights in the front of the house, presumably associated with the UFO. Looking out the door, she observed a hand reaching in from outside, long and skinny. It appeared to be very low, the entity about the height of a small child. At this point, she also remembers a nonverbal scolding, "you know better than to try to resist," telepathic thoughts rather than words. At that point, she remembers nothing more of the encounter.

Her next memory was of standing in the kitchen in her nightgown, looking through the east-facing window. Although she could not remember it, she felt as if, in some way, she had just come in through that window. Outside, she observed a bright light ascending above and slightly to the east of the house. It shone through the window, projecting a pool of light onto the kitchen floor. Then her memory again discontinued. The next thing she knew, it was morning.

In many ways, Evelyn's account is typical of the consciously remembered part of many close encounters. Characteristic of such events is the high degree of strangeness. There is deepening interaction with the phenomenon, including a close UFO sighting and communication with the object or entities. Most importantly, events of the fourth kind often include missing or incomplete recall; the sense that far more happened than one remembers.

There is an accompanying sense of mystery, which can often include reality paradoxes and inexplicable thoughts and emotions. We will discuss these strange aspects of the close encounter in the next chapter, as we examine the UFO abduction in detail.

UFOs and Close Encounters in Perspective

Nearly ten percent of humanity has had sightings of the unexplained. To perhaps as many as one person in a hundred, the phenomenon has become real and deeply personal. Given the number of witnesses and experiencers, my sense is that those whom UFO researchers have interviewed are only the tip of the iceberg. The examples in this chapter are just a few of the many sightings and encounters in our case files. For every UFO encounter known, many more probably remain unreported.

Such sightings and encounters suggest the depth of the relationship between witnesses and the UFO phenomenon. They hint at deepening meaningfulness as the strangeness of the encounter increases.

What occurs during the CE4 event, itself? What can we learn when we begin to study contact with the UFO phenomenon as an overall process rather than simply as individual events? In the next chapter, we examine in more detail the close encounter of the fourth kind. In subsequent chapters, we will examine what may be behind the encounters and what the intention of the visitors might be. As we further examine this strange influence in our world, we will begin to see an emerging process of contact that may be far more coherent than we thought.

Chapter 2 – The Enigma of the Close Encounter

It is late at night and the lonely road stretches ahead. That light in the distance, the light that seemed different from anything familiar, is now approaching. As it does, you begin to realize that you are the object of its attention.

Then it arrives.

You see a presence, beings emerging from within the object. Now you realize that, for better or for worse, your UFO sighting has become a close encounter. The UFO phenomenon has now become your reality. For you, the mystery has taken on a new depth...

In the previous chapter, we discussed UFO sightings, cases in which the witness remained in one location and the UFO in another. We also noted that, to varying degrees, we can think of a sighting as a relationship between witness and phenomenon. Yet, during a sighting, even if the object is relatively close, the relationship may still be comfortably distant.

What happens when the phenomenon begins to interact overtly with the witness, when the relationship becomes up close and personal? What happens when one finds oneself experiencing something far beyond a sighting, when the witness becomes an experiencer of the extraordinary?

The Close Encounter Reality

As the level of strangeness and meaningfulness of a UFO event increases, we begin to enter the realm of contact, the close encounter of the fourth kind (CE4). In many ways the realm of the phenomenon, be it the interior of an alien spacecraft or some other dimension, can be thought of as an alternate reality to that of the experiencer's normal world. In a close encounter of the fourth kind, a reality transformation takes place and the witness becomes part of that phenomenal realm. Willingly or otherwise, he/she has become an experiencer, a full participant in the UFO phenomenon.

There are many types of encounter. Some experiencers describe events in which they have been contacted, often as an equal with the visitors. It is a positive, open, often enlightening experience. For such experiencers, whom we will refer to as contactees, an event is nearly always remembered in a favorable way.

Unfortunately, the visitor encounter more commonly takes the form of UFO abduction. For abductees, those experiencers undergoing such involuntary events, the visitor interaction takes a far more intrusive form. In this chapter, we consider this class of encounters. We will examine different aspects of the abduction phenomenon. In so doing, we will see how "standard" themes begin to emerge within the accounts of UFO encounters.

Evelyn: The "Pulled out of Bed" CE4

In the previous chapter, we met an experiencer we called "Evelyn," who has had a lifetime of close encounters. Probably the most powerful of these was a UFO abduction that left a legacy of fear in her life. This event occurred one chilly late-fall night in northern Wisconsin. Snow had not yet fallen, but Evelyn and her family had been busy that day preparing their large rural yard for the coming winter.

As Evelyn lay asleep in bed, she was awakened by a sudden touch on the shoulder. She felt a presence at the side of the bed next to the wall. A moment later, she felt something or someone grab her feet. She instantly felt herself floating, pulled toward the foot of the bed. In panic, she tried to scream and kick but found herself unable to do so. While being pulled toward the end of the bed, she dragged her arms behind her, unsuccessfully trying to awaken her husband.

Evelyn's body had now floated partially off the bed. From waist to feet, she had floated beyond the edge, with her upper body still suspended over the bed. At that point, her memory of the event abruptly discontinued.

Some time later, Evelyn awoke in fright, a fear that would continue to haunt her for years. In the following days, the family discovered a large newly formed burned spot in the back yard, implying that this had not been simply a bad dream or imagination.

Rather, it had been a real, physical event. It would forever alter the lives of Evelyn and her family. Only with the help of hypnosis, a couple of years later, could Evelyn recall these events, revealing an encounter in the alien realm.

During hypnotic regression, Evelyn recalled floating out of the bed and passing horizontally through the window, accompanied by small alien beings. Next, she recalled standing upright, yet somehow floating through the back yard toward the UFO. She then ascended an incline into the interior of the object.

Within the object, Evelyn found herself in a curved hallway filled with bright light. Several beings were present, both alien and human appearing. Then she recalled being in a large, bright room, where she apparently underwent some form of medical-related procedure.[3] Subsequently, she remembered being back in the hallway, interacting further with the occupants of the craft.

Later, as her encounter concluded, she found herself escorted back to her house and then to her own bed. The next thing she remembered, it was morning. Even though she remembered nothing of the encounter events, she awoke with a new, intense feeling of fear.

Later on, we will learn more about Evelyn's encounters - strange and powerful events that forever changed her life.

---

[3] Note: Due to the limited session time, and the fear this memory generated, we were not able to explore this portion of the encounter in more detail.

Janice: Wolves on the House

Another event occurred to the experiencer named "Janice," whom we also met in the previous chapter. During the wee hours of a July night in the early 1990s, she suddenly found herself standing in her front yard in her nightclothes and bare feet. Looking toward her house, which was about 100 feet away, she could see what looked like four or five "coyotes or wolves." These appeared to be sitting on the top of the roof staring down at her.

She was not able to observe them in detail, only sensing that they looked "furry." She felt they were looking at her, yet she could not recall their eyes in detail. Both curious and frightened, she slowly began to walk toward the house. As she did, she found herself escorted by two entities. Both were slightly behind her, one to either side. She felt herself guided by them and suddenly felt as though she were being lifted, weightless and rising. She felt a "sick" feeling, as one might have in free fall.

Janice then found herself in a white, brightly lighted hallway. Instructed to walk down the hallway, she felt mentally compelled to comply, moving toward a doorway approximately 10 feet to her right.

She walked through the doorway and entered a room with a large machine on the right side. A smaller entity stood at the far end of the room to the left of the machine. Janice was initially very interested in the machine, which to her seemed like a "fascinating piece of technology." It had a large number of bright, multicolored lights on it. She was briefly allowed to examine the machine but as she did, she suddenly found herself a few feet to the left, in front of the middle portion.

The middle portion did not appear to have the lights and buttons, which were present on the right-most part, but appeared to be an open section. She observed what seemed like a one to three inch thick flow of "green goo" flowing from the right-most part of the machine, through a trough that crossed the lower middle portion, and disappearing into the left section. She was instructed to put her hand in the "goo" which she found "gross and repulsive."

She initially resisted, but was again mentally compelled to follow her instructions.

She felt that the machine was an experiment or game, whose sole purpose was to interact with her. When she placed her hand in the substance, contrary to her expectations, it "did not feel icky." She felt no resistance, as if she were "putting her hand into air." To her relief, instead of some form of 'goo', it seemed more like a vapor, having little or no substance. Rather, it seemed to "merge with her hand."

Her memories then suddenly shift from her encounter with the "goo machine" to a conference with a number of entities in a larger circular room. The room seemed featureless, with no visible doors or corners. She was accompanied by approximately five entities. She had a full view of the beings and could see their features in detail. The larger entity was about five feet tall and the other entities were considerably shorter. They were gray in color and wore tight fitting uniform-like clothes.

The larger entity appeared to be mystified and asked her a question that pertained to her health at the time of the event. She gave a brief explanation to the tall entity, which then turned around and repeated this to the smaller entities. They then repeated it in what the witness felt was unison. Janice felt that all communication was non-verbal, probably telepathic.

Following this exchange, Janice immediately found herself waking up back in bed. She sat up suddenly, thinking, "What was that?"

This experience and others like it have been a thread of mystery in Janice's life. For personal reasons Janice has not explored this experience with hypnotic regression. Thus, many more details may be present but unexplored. Yet with conscious memory alone, Janice has recalled many memories of her experiences in the alien realm.

The Close Encounter Prototype

We have seen how, during experiences such as those of Evelyn and Janice, the relationship with the phenomenon is transformed from that of a sighting witness into that of an active participant in the encounter. The relationship deepens and the experiencer and entities become fully interacting players in the cosmic drama of the close encounter.

Experiences such as those described above are typical of reports from the many abductees whom I and other researchers have worked with. In all cases, the witness is somehow captured and transported into the alien realm.

The literature and our case files contain many such descriptions and from these, a set of common themes emerges. These have become a well documented; though still mysterious, aspect of the UFO abduction phenomenon.

Examples from the Literature

In his book *Intruders*, Budd Hopkins describes a series of UFO abductions occurring in a community that he gives the pseudonym "Copley Woods." The book centers on the experiences of an abductee whom he refers to as "Kathie Davis" and his investigation of these events. It describes a series of abductions in which medical, reproductive and genetic procedures become evident. Within the pages of his book emerges an account of how Kathie was repeatedly abducted and was impregnated with some form of genetically altered or hybridized embryo. The book describes how, during an encounter two to three months later, the partially developed fetus was removed. Some time later, in yet another abduction event, Kathie was shown an alien appearing, but also very human, child.

Budd Hopkins develops the case that this has happened to uncounted numbers of other abductees and that a systematic process of abduction and medical procedures seems to be occurring. He builds the case that the medical procedures focus on human reproduction and genetics and that the aliens appear to be creating some form of human-alien hybrid.

While closely following the case of Kathie Davis, he describes several other encounters and notes the degree of similarity of these cases. Thus, Hopkins was among the first to provide a focus on this type of encounter, a prototype that seems to be typical (or standard) among UFO encounter reports. I therefore refer to this as the "Standard Scenario" or "Standard Model" of UFO abduction.

In his book *The Threat,*[iv] David Jacobs develops the theme of genetics/hybridization to an even greater degree. He has cataloged an extensive collection of abduction cases, including many sessions of hypnotic regression with abductees. Jacobs describes genetic tampering and hybridization similar to that described in *Intruders*. However, in *The Threat,* Jacobs focuses upon what he thinks may be the intent of the visitors. What emerges in his book is an alien agenda directed toward a sinister but unknown purpose.

The basic encounter scenario Jacobs describes appears to be nearly identical to that described by Hopkins. It includes capture and abduction by small gray aliens, medical/reproductive/genetic procedures and the birth of apparently genetically modified or crossbred children.

In his book *The Watchers,*[v] Raymond Fowler develops a detailed chronology of the close encounter events of the Andreasson-Luca family. This book also describes in detail, a pattern similar to that described in *Intruders*, *The Threat* and other works. Like Jacobs, Fowler further develops the abduction-pattern theme. Although he sees a different purpose to the phenomenon, he cites the same common sequence of events - that of abduction, medical experimentation and hybridization. Thus Fowler, like most other authors, captures the standard encounter scenario and its powerful influence on the life of the experiencer.

The Close Encounter Scenario

While experiencers may relate a variety of descriptions of their encounters, we indicated earlier that abduction accounts often have a basic core of similarity. Taken in aggregate, they form the typical scenario of UFO encounters, initially framed by Budd Hopkins, David Jacobs, Raymond Fowler and others.[vi][vii] Earlier, I referred to this as the "Standard Model" of UFO abduction.

A standard event generally includes the capture of an abductee, typically from their bedroom or while in their car. The abductee is later returned to his/her original location, with one to two hours apparently missing from their lives.

In many abduction events, the experiencer may remember only the beginning and end of the encounter with no memory of what took place in between. However, in some cases, partial memory of the encounter may exist.

As we examine the events, either spontaneously recalled or using a technique such as hypnotic regression, we find that they tend to follow a pattern.[viii]

1. There is attention diversion toward the phenomenon. This can include the close-up sighting of a UFO[ix], a change in the environment (such as sudden quietness or sense of unreality), or a change in state of mind (such as the compulsion to pull the car over to the side of the road, or an inexplicable feeling that you "must go outside").

    A spectacular and classic case of attention-diversion occurred in Minnesota in the 1980s, when two friends were traveling from St. Paul, Minnesota, to a location in Wisconsin. While driving east on I94 from St. Paul, one noticed a light, low and to their right. To both of them, the light looked unusual. Their next memory is of having turned off the interstate onto a gravel road. The road led them directly beneath the light, which they observed for a matter of minutes as it maneuvered directly above them. When they arrived at their destination in Wisconsin, they realized that they were at least forty-five minutes late.

2. Some form of capture scenario ensues. The abductee is somehow immobilized. Entities appear around the abductee, who is then escorted to a waiting UFO. As in the case of Evelyn and Janice, described above, the experiencer often has partial memory of these early events.

3. The abductee suddenly finds him/herself within the domain of the phenomenon. Often, this is in a brightly lit room aboard the UFO, with the abductee sitting or lying on a metallic examining table.

4. The abductee undergoes some form of medical examination or scientific experiment. There is often a sexual or reproductive aspect to this procedure, such as the harvesting of sperm or eggs. On occasion this may involve the implantation or removal of an embryo – an aspect of the phenomenon that is at best, extremely traumatic to the abductee. We will have much more to say about this aspect of alien encounters in later chapters.

5. Additional interaction may occur, such as the teaching of some sort of lesson or a telepathic dialog between the abductee and entities. Many abductees have described seeing parts of the interior of the UFO. Others describe being shown or briefed on future events. Many describe being given special knowledge and are often told that they have a mission to carry out.

6. The abductee is returned to his/her original location, usually after approximately two hours, with little or no memory of that time. To the rational mind, the remembered events often do not make sense. Thus, on return from the encounter, the abductee often feels a confusing sense of discontinuity or ambiguity.

Researchers have noted how these events form the core of the close encounter experience. As mentioned above, the conscious memory of the UFO encounter is often limited to the capture and return steps. Yet, while unavailable to the conscious mind, memories of the remaining portions of the encounter often remain within the subconscious. Under the right conditions, the "interior scenario" may emerge, often when the experiencer undergoes hypnotic regression.

For Evelyn and many others with whom I and other researchers have worked, the consciously remembered portions of their experiences are only the tip of the iceberg. Under hypnosis, many additional details have emerged. Often, these match the "standard" scenario previously described. We will more closely examine some of these events later, in this and subsequent chapters.

Additional Commonalities

In addition to the abduction scenario described above, researchers have noted an extensive list of themes that further seem to comprise the close encounter.

1. The entities at the core of the phenomenon are usually those commonly referred to as "grays." They are typically small beings, about three feet tall, with a larger gray being, about five feet tall, acting as leader. Nearly all abduction experiencers describe these as little beings with large black eyes. They are a nearly ubiquitous aspect of the abduction experience.

2. Nearly all abductees describe clairaudient telepathy as the means of communication within the realm of the phenomenon. Most describe the "speech" of the entities as sounding much like a disembodied voice in a language the abductee understands and which they frequently describe as being "inside my head."

3. Many describe a process, which David Jacobs has labeled "Mindscan," in which an entity with large black eyes stares deeply into the eyes of the abductee. Because of this, the entity appears to make intimate mental contact with the psyche of the abductee, reading his/her thoughts and perhaps implanting information or mental scenarios.

4. Scenarios of disaster, apocalypse and related warnings are often telepathically "beamed" into the mind of the experiencer. More often than not, these constitute a warning about the severe consequences of war and/or human mismanagement of Earth's environment. One experiencer, whom I refer to as "Carolyn," has had experienced many such images, which we will describe later in this chapter.

5. Entities often appear to the abductee as animals, cartoon characters, deceased relatives, or other relatively familiar forms. Researchers often refer to these illusions as screen memories. Multiple sightings over a witness's lifetime of strange animals or other, similar anomalies may be indications of past encounters.

One apparent screen memory was described to me by an experiencer, whom I will call "Kimberly," who lived with her husband and two daughters on a farm in Wisconsin. Kim has described many strange events occurring both to her and to her daughters. One event occurred early in the morning as her daughters were about to walk from the house to the barn. In their back yard, they observed one or more "wolves" that appeared to have large black eyes. When the girls spotted them, the wolves suddenly moved from the yard, passing through (not over, but through) the fence. They then disappeared into the woods.

Another such memory, described by Evelyn, occurred as she was about to leave her house for work. Evelyn went to work very early each morning - usually well before sunrise. On this particular morning she had gotten into her car and was about to back into her driveway when a large "owl," several feet tall, appeared on the hood of her car. Like Kimberly, Evelyn also observed the large black eyes common in such memories.

6. Close encounter experiences often tend to follow along family lines. Many experiencers, such as Evelyn, describe how their children have also had encounters and a lifelong relationship with the visitors (we will discuss this lifelong relationship in the next chapter). In his book, *Intruders*,[x] Budd Hopkins first notes this common theme, describing the encounters of Kathie Davis and her family. In *The Watchers*, Raymond Fowler describes a similar family lineage within the Andreasson-Luca family.

These additional aspects provide a further thread of both consistency and mystery within the close encounter experience. Janice, Evelyn and most other experiencers in this book, as well as nearly all abduction experiencers in the UFO literature, extensively describe these classic aspects of their experiences.

Post Encounter Effects

As we stated earlier, the details of the events that occur during an abduction often remain unremembered at the conscious level. Yet, knowledge and memories of close encounters may linger just below the experiencer's consciousness as strange half-recalled events, with a powerful aura of mystery or strong emotion. This imprint in the subconscious can have a powerful effect in the life of the experiencer.

For Evelyn, the months following her encounters were filled with fear and mystery. Sounds in the house became sources of concern. She became both fascinated and fearful about the subjects of UFOs, extraterrestrial life, space and related topics. The night sky itself became a source of both fascination and fear.

In addition, paranormal effects began to manifest in the house. Light bulbs would burn out with annoying regularity. Watches and clocks would speed up, slow down, or simply stop.[4]

In desperation, Evelyn and her family began to reach out to others they felt they could trust, including counselors and their minister. Eventually their minister put them in contact with someone who was able to direct them to the Mutual UFO Network (MUFON). Thus, our work began with her.

Hypnotherapy helped to resolve her fear and uncover the incredible events of that night. During hypnotic regression, Evelyn recalled being taken aboard a UFO that had landed near the house. Once on board, an extensive interaction with the visitors occurred including a medical examination and telepathic dialogs with several types of aliens. As a result of bringing these details to light, she was able to come to terms with the fear originating from that night.

---

[4] Paranormal effects such as these often occur to experiencers in the days to weeks following UFO events. Researchers often refer to this "paranormal fallout." Like many aspects of the UFO encounter phenomenon, these effects remain an unsolved mystery.

For many experiencers, post-encounter effects can take the form of flashbacks, intrusive thoughts, or memories. For one experiencer, whom I refer to as "Carolyn," these took the form of obscure thoughts of apocalypse and memories that seemed to have little or no context in her life. Her memories were of hospital-like buildings with which she was unfamiliar and of bleak scenarios in which much of the known world had been destroyed. Her memories suggested that she would play a role in such a world, though that role remained ambiguous.

For many such as Carolyn, post-encounter effects include both this sense of apocalyptic warning and this feeling that they have a special role or mission. Many experiencers describe how they have some duty to perform in what they call the "Coming Changes."

Another experiencer, whom I refer to as "Jenny" and describe later in this book, has described a mission, given to her during her interaction with the visitors, to become a healer. Through her healing work, she was instructed to spread a message from the visitors throughout humanity. This mission/message aspect of the phenomenon is described extensively in *The Threat* and other abduction literature. It seems to be a core aspect of the phenomenon.

For Evelyn, Carolyn, Jenny and so many other experiencers, unexpected and uninvited paranormal effects, fear, apocalyptic visions and the sense of message/mission, often mark everyday life. In the next chapter, we will explore how these and other aspects of the phenomenon can be powerful agents for personal transformation.

Outward Signs of Unremembered Encounters

As we indicated earlier, the events of a close encounter frequently remain unremembered. Yet, as we have noted, there are often outward signs to suggest that such events have occurred in a person's life.

In their book, *Healing Shattered Reality*[xi], Alice Bryant and Linda Seebach describe a number of indicators which, when taken in aggregate, strongly suggest that a person may be having close encounter experiences. The following list is condensed from Bryant and Seebach's excellent work, delineating some of these outward signs:

1. A sudden and perhaps inexplicable shift in interest toward topics related to UFOs, Extraterrestrial life, space and space travel, the universe, eastern spirituality or related topics,

2. A shift in perspective or beliefs to a holistic, metaphysical, environmental and pacifist viewpoint, along with a sense of citizenship in the world - and even of the cosmos,

3. A sudden sense of mission, life-purpose, and/or the feeling of possessing secret knowledge, perhaps something to be revealed when the time is right,

4. Nightmares, dreams and/or memory fragments related to UFOs or non-human entities,

5. An awakening of psychic abilities such as telepathy, clairaudience, clairvoyance, or psychokinesis,

6. Minor physical injuries such as bumps, bruises, scrapes, puncture marks or scoop marks, for which they cannot determine a cause. These often seem to recur inexplicably while the experiencer is asleep.

7. Memories or experiences of visits-by or sightings-of ghosts, monsters, religious figures or other extra-normal entities,

8. Multiple meaningful UFO sightings throughout life,

9. Missing or accelerated time: for example, an auto trip that seems to take too long or in which arrival seems to occur too fast.

When several of these signs are present, they strongly suggest the presence of the phenomenon in a person's life. For experiencers such as Janice, Evelyn, Jenny, Carolyn and many others, these form a prominent theme in their lives. For many others as well, they suggest that something - an undercurrent of the visitor experience - is present within their reality.

In public presentations on the close encounter phenomenon, I often describe these outward indications and take a brief, informal poll of the number of people who have experienced one or more of them. The number of people who have indicated experiencing such effects is nearly one person out of 100. Most often, in subsequent interviews with people who have described such indicators, I have noted that the presence of the visitor phenomenon is apparent in their lives. While no single indicator is an absolute marker of the phenomenon, when taken in total they tend to be reliable clues to the presence of the unexplained.

A Lifetime of Contact

At its simplest, a sighting or distant encounter may be a one-time-only event in a person's life. However, a close encounter experience is usually one event in a lifelong sequence. Probably, we can best understand it as a long-term process of contact between the individual and the visitors. In my case files and those of other researchers, it appears that many UFO experiencers seem to have this lifelong "dance" with the phenomenon. From distant sightings to close encounters, their lives seem to contain this long chain of interplay. Evelyn, Janice and others have described multiple experiences, extending across all levels of strangeness and meaningfulness, occurring throughout their lives.

In his book, *The Watchers*, Raymond Fowler extensively describes the interactions of the Andreasson Luca family with the phenomenon. In addition, in their books, Budd Hopkins and David Jacobs also portray the phenomenon as a life-long interaction.

David Jacobs suggests that its onset occurs at about the age of three. In a similar manner, nearly every experiencer has told me that their first event occurred during their early childhood. Abductions then continue on an infrequent basis throughout their early years. I have noted that many experiencers tend to remember this childhood interaction as a playful, magical influence in their lives.

The literature continues to describe how, as abductees near adulthood; their experiences become more serious in nature and begin to take on a sexual overtone. Medical exams occur during these abduction experiences, with a heavy emphasis on reproduction. At this point, many experiencers describe the harvesting of sperm or eggs.

As mentioned earlier, many female experiencers describe being impregnated as a result of their encounters.[xii] This pregnancy seems to last only a few months, after which time they are re-abducted and the partially developed fetus removed. In subsequent abductions, they may be shown a developing baby contained in some type of incubation device, usually a large vertical, fluid-filled "test tube." A few years later, during another abduction, the experiencer may be shown a baby or child, frequently very alien looking, that is now living on its own within the alien realm. They instinctively sense that this is their child. We discuss this aspect of the encounter phenomenon in more detail in chapter 6.

At some point in their lifetime of interaction with the phenomenon, many experiencers also develop enhanced psychic abilities and may suddenly find themselves mentally or spiritually transformed. We will examine these transformational and awakening aspects in chapter 4 as we discuss the human side of close encounters.

Conclusion - The Close Encounter Enigma

In this chapter, we have merely scratched the surface of the enigma that is the close encounter phenomenon. However, we have seen a core of commonality to it that pervades the life of the experiencer. We have also seen how encounters are related to human reproduction and genetics, as well as human (para)psychology and the (potentially apocalyptic) future of humanity.

In the next chapters, we will examine high strangeness cases of close encounters and the impact that the encounter phenomenon has on the life of the experiencer. In subsequent chapters, we will examine the fundamental "big-picture" questions regarding the close encounter phenomenon, including speculations on its possible motivations and objectives. Throughout, we will continue to develop a picture of the relationship between the experiencer and the phenomenon – the Cosmic Bridge between ordinary life and the unknown that is the realm of the close encounter enigma.

Chapter 3 - Extraordinary Encounters: Beyond the Standard Model

The close encounter phenomenon presents us with a theatre of the strange. It stretches our personal reality and challenges our beliefs about our place in the universe. Yet in the last chapter, we saw how, even though far beyond the normal paradigms of daily life, many such encounters seem to form a pattern. This, we called the "Standard Model" of alien abduction.

However, even though we find this common thread of events, accounts from many experiencers suggest that the "standard" encounter is by no means the only scenario. Many details go far beyond it. In my own work with experiencers, I have found a considerable variety of accounts. For each classic abduction event, I often hear one or more other accounts that portray very different scenarios.

At some times in the life of an experiencer encounter events may bear greater resemblance to the "standard" scenario. Yet at other times, their encounters may be very different. The variety and strangeness of such encounters bring us even more paradox. While they deepen the mystery, they also offer us important keys to understanding.

In this chapter, we will examine some extraordinary encounters. We will see how, for the experiencer, such cases further blur the boundary between the everyday and extraordinary realms - a Cosmic Bridge between worlds.

Details that "Stretch" the Standard Model

Although strange beyond imagining, nearly all experiencers describe events that seem to conform to the scenario as described in the previous chapter. Yet often, as we look more closely we can note how an abduction event may be different. The degree of difference may vary but even within a relatively "standard" encounter, there may be varying details. While some may vary only slightly, we will see in this chapter how others are very different, indeed. Thus, within this already-strange phenomenon we see a range of cases, a further spectrum of strangeness.

Jade: A Remembered Encounter

One abductee has described an event that clearly illustrates this "slightly different" type of encounter. This event took place during a mid evening in November of 1990. The witness, who has adopted the pseudonym of "Jade," left the front door his house following an argument with his family and began to walk toward the house of a friend who lived in his neighborhood. There had been a recent snowstorm leaving the streets covered with approximately 1/2 inch of snow. He was wearing light shoes and a light jacket. His shoes quickly became wet and after a few minutes, he became increasingly chilled.

Approximately half the distance from home, he became disoriented. Originally walking west while looking down at the ground, he suddenly looked up and realized that his surroundings were unfamiliar to him. He determined that he had somehow reversed direction and was now heading east. He also noted that he now felt warm, as if he had just emerged from a warm house or car. Further, his previous emotional upset had vanished and he felt in a much more positive state of mind.

Flashbacks and Dream Recalls:

Within a year after this event, Jade began to have flashbacks in which he remembered seeing a large object (described as a dark circular silhouette) over him with a brilliant light shining down on him. Following the initial flashbacks, he began to remember more and more of the events with increasing detail. This included dreams and visions of objects [presumably associated with the event]. He began to have flashbacks with high enough frequency that he kept a sketchbook of them. With each dream, the details became clearer. Memories of places and things, presumably related to the encounter, started coming back to him. The following description is a composite scenario derived from them.

Initial Encounter Scenario:

A scenario of his initial encounter emerges in which he was walking slowly along the street, looking at the ground. He was getting quite cold and as he walked, he tried not to think about it. There was no discernable wind and snow had just fallen. Cars would occasionally drive by and he could hear the crunch of snow on the tires. The road had not yet been plowed. Thus, he was trying not to get snow on top of shoes, which would otherwise soak through the leather.

There was flash of light (in subsequent flashbacks this was a spotlight-like pool of light) on the snow. He recalls no sound of any kind but remembers that the light appeared to come from above. It appeared elliptical in shape, like a spotlight at an angle. He looked up to see where the light was coming from and saw a large object slightly to his north, just beyond several nearby houses.

He remembered seeing the object at treetop level and as he watched, it moved straight overhead with the light staying on him like a spotlight. Earlier, when it had been more to one side, he had felt that the light was not as bright. But when it was directly overhead it became blinding. He estimated the beam to be about 6 feet wide and the diameter of the object to be about 20 to 30 feet.

He covered his eyes and looked down. Then, putting his right hand over his face, leaving just a little corner to see through, he again looked up at the object. He was able to see the dark edge of a large circular object with the light coming from the middle. Due to the brilliance of the light, he was unable to see anything except the edge.

Interior Scenario:
From subsequent flashbacks, a picture of the apparent interior of the object emerges. The memories begin with being awake, presumably inside the object. Jade was apparently alert but unable to move. There was someone behind him, gently pulling him across the floor into a dark room of some kind. He was being "almost dragged" backwards, as if someone was carrying him from around the chest. He remembers being moved about three or four feet and then placed on a platform, apparently a table, rising from the floor. He describes the table as cold and hard.

He lay on his right side on the table, with an entities standing on either side of the table, one in front of him and another behind. All conversation appeared to be non-verbal (apparently telepathic), with the speech appearing to occur within his own head. He noted them appearing to disagree over something.

The entities all looked similar. He could clearly remember their eyes, which appeared to be associated with their communication. The entity on his right was "looking into" Jade, apparently communicating with him. He sensed some form of a relationship with the entities and, even though they were strangers, they appeared to have "sympathy" for him.

The entities seldom if ever turned away from him, so Jade's view of the entities was largely from the front. At the level of the table he was able to see the entities from about the middle of their chest on up. He noted that their bodies seemed to have approximately human proportions.

The entities had large heads with small chins. Their faces were "flesh" colored, with a dull yellowish tone, but seemed "hollow," almost like a mask. Their cheekbones were more prominent than what he had seen in drawings of a "typical" gray. He did not remember seeing a mouth. The hands appeared to have two thumbs [not sure how many fingers]. They did not have any hair.

The back of the head flared out beginning at the temples. He also remembered the eyes being spaced in about human proportions and being somewhat "oriental" looking. They did have whites, but the pupils were extremely large. The entities were wearing what appeared to be "jump suits." Jade noticed a vee-shaped collar on at least one of the entities.

Jade remembered the room as being round with pillars around the outer edge. The ceiling of the room appeared to have a diameter greater than that of the room. There was a gap between the top of the wall and the ceiling. The ceiling was domed with the light and apparatus coming down from the middle. At the edge of the circular room, there was a rounded wall with regular pillars.

Unique Elements in Jade's Encounter:

In the encounter that Jade describes, we find nearly all of the elements of the standard abduction scenario. However, there are some fascinating variations. The most important of these appears to be the description of the entities themselves. They do not seem to have the gray-like features so commonly described by abductees. In addition, they do not seem to exhibit the hierarchical leader-worker structure so common in descriptions of the grays, where the taller gray is the boss and the small grays are minions.

It is not known if this is one of many encounters or if this encounter was an isolated event in Jade's life. However, all of Jade's memories appear to pertain to this one event.[5]

In addition, the fundamental nature of Jade's abduction seemed to have a different characteristic than do others. While memory may only be partial of this event, there does not seem to

---

[5] In subsequent conversation with Jade, he indicates that he has maintained an interest in UFOs, but feels that he has apparently not had additional close encounter experiences. In addition, as I talked with him in preparation for this book, he expressed a degree of skepticism regarding the reality of his encounter memories.

be a sexual overtone to it. The encounter appears more related to healing than to reproduction, with Jade's emotional (and possibly physical) needs being addressed.

Evelyn: Metaphysical Events during a CE4

Evelyn, introduced earlier, has had many encounters during her life. On the surface, many elements of her encounters appear to fit the standard model. Yet her encounters also have had many unusual aspects, which do not fit any of the prototypes of the typical encounter.

During hypnotic regression, Evelyn described many details of her "pulled out of bed" and her "laundry room" encounter (see Chapter1) as well as other events. In each case, she told of being surrounded by small alien entities, closely matching the common descriptions of grays. She was then taken aboard a UFO. She described the near-total amnesia surrounding these events, a phenomenon nearly universal among abduction events.

Many key events were indeed very similar to those so commonly described in classic abduction accounts. Yet, mixed in with these descriptions are details that vary considerably from the standard model of UFO encounters. In Evelyn's case, there was the significant inclusion of spiritual and metaphysical events within her physical close encounter experiences.

A few months before her key encounters began she experienced an encounter with a metaphysical "dream" entity, which she referred to as "Marcus." She described having a positive relationship with this being, which subsequently appeared in the context of her encounter experiences.

The entity, which she described as humanlike but with some significant variations, did not resemble any I have noted in the close encounter literature to date. In addition, her positive relationship with the entity seemed in sharp contrast to the traumatic nature of the subsequent abduction events.

As we explored her relationship with Marcus, it became evident that he was more metaphysical in nature. He had appeared about a year before the beginning of her key encounters (described in the last chapter). In this event, he had appeared in a field or bubble of light over her bed. He held out his hand, which she took. She then found herself within this field of light. An interaction occurred which she described as a "dance."

Evelyn described Marcus as her companion in another time and place, apparently another lifetime. His association with her abduction experiences initially seems tenuous. Yet in another way, Marcus was inextricably bound to those same events. His initial appearance seemed to herald the beginning of her key encounters. She also noted that he was present at least once when she was returned from an encounter.

While Marcus appeared in some way to be metaphysical in nature, he also seemed to provide support for Evelyn during her difficult abduction experiences. The relationship between this entity and Evelyn's other encounter experiences appears to be well established, but does not seem to fit the "standard" encounter model. Thus, while Evelyn's experiences have many of the "standard model" elements in them, the presence of Marcus and related events also seem to support a more metaphysical understanding of the phenomenon.

Cases that defy the standard model

While to some degree, many encounter descriptions seem to fit the standard model of UFO abduction, we also find a large number of mysterious events that have no similarity to this common mold. These are what I loosely describe as "nonstandard" encounters and they suggest to me how little we really know about close encounters.

In this section, we examine several of these encounters, taken from my own case files. We then consider some possible implications of these events.

"Lisa" and "Mike": Two Non-Standard Encounters

A number of close encounter events, typified by those of an experiencer whom I will call "Lisa," show many characteristics which vary considerably from the "standard model." In one key event, Lisa remembered herself and several other people in a car, which was being approached by a UFO. They were taken as a group into a "scout craft," which then took them to a larger ship where they were met by human-like entities. The entities appeared to be wearing military uniforms from an earlier time, which Lisa thought might have been the 1940s. They conferred with her on a military struggle in which they were apparently engaged and in which they said that she had sided with them. However, she felt suspicious of their intentions and resisted their questions.

Lisa and the other abductees in his group were then taken to a "base" where Lisa was met by another military leader, also human appearing. Lisa then remembered being placed back in their car[6] and apparently returned to their earthly reality.

---

[6] Note: It is not clear if the abduction was from a car or if the car, itself, was somehow part of the abduction scenario. Perhaps the car, itself, was a screen memory for another aspect of the encounter.

Another such "non-standard" event was one which occurred to an experiencer, whom I will call "Mike." Late at night, he felt the sense of presence and paralysis common in the beginning of encounters. He then observed entities appear through a nearby mirror. The entities indicated that they were from another galaxy (Andromeda). They then escorted him back to their realm and Mike found himself in a vast alien domain, which looked both paradise-like and technological. The entities (looking completely unlike "standard" types of aliens) escorted him to meet a leader figure who interrogated him. The leader informed him of a mission which they had planned for him. They then returned him to the normal human world and back to bed.

Ambiguity and Paradox

Neither Lisa nor Mike, nor other similar experiencers, have described their experiences as good or evil. While almost none felt their encounters were dreams or metaphysical in nature, these experiences seem very different from classic "standard model" encounters. Yet they do tend to contain a few "standard" elements such as paralysis at the experience onset and accepting a mission or a role in coming events. This varying level of standard-ness seems to be typical in the anomaly "careers" of many experiencers. Some events fit the mold while others do not.

The examples described above only scratch the surface of the variety, ambiguity and paradox inherent in the UFO and close encounter phenomenon. Often encounters are dreamlike in nature, having an "illogical" character to them. Yet often, as in Evelyn's experiences of the previous chapter and similar cases in the literature, the phenomenon also leaves physical evidence; such as marks on the body, ground traces, etc., concrete manifestations in the lives of the experiencer. Thus, these descriptions suggest that rather than being totally nuts-and-bolts or totally dream-like, in some way they are both simultaneously. This suggests a degree of plasticity to the reality of the phenomenon, perhaps a superposition of multiple levels of reality within the experiencer's consciousness

In the cases of Mike and Lisa, their experiences bear little resemblance to the standard model. Yet for both they are part of a lifetime of encounters that also include "standard" type events. For both of them the phenomenon seems to have a deeper dimension than a simple "nuts and bolts" model of close encounters. In both cases, the phenomenon seems to have played a key role in their lives, with profound physical, mental and spiritual effects.

Extraordinary Phenomena in Contact

So far, we have glimpsed the world of the close encounter and seen how the reality of the witness shifts to that of the phenomenon. We have explored the events on the far side of this shift and discovered a spectrum of mystery. Yet we have suggested that there are further elements of close encounters events which we have not yet explored - the strangest of the strange.

One of the best ways to probe any phenomenon is to study it in cases where it stretches existing theories to the breaking point. Even within the close encounter phenomenon, extraordinary in its own right, there are even stranger elements. Thus, perhaps one way to better understand UFOs and close encounters is to look at cases of high strangeness.

The close encounter phenomenon has many truly bizarre aspects to it. Experiencers often describe events such as the Oz factor, in which reality itself appears to change during an encounter, spontaneous invisibility, and related phenomena. These are just some of the effects that challenge our understanding of the physical world. In this section, we will examine some of these bizarre aspects of an already deep mystery.

Reality Anomalies

In many cases, the first impression of listeners, upon hearing descriptions of a close encounter, is that "it must have been a dream." Researchers have long noted that the dream-like qualities of CE4 narratives suggest a possible non-physical nature to such encounters.

Yet frequently there are also well-defined physical aspects to these events. Some CE4 cases have been associated with well-defined UFO sightings. At times, the UFO has been observed by multiple witnesses, or seen leaving physical traces. Cases of this nature are well documented in the literature,[xiii] especially in books by Budd Hopkins such as *Missing Time, Intruders,* and *Witnessed.* In many cases, both the physical and metaphysical views appear to be valid simultaneously[xiv] - a superposition of realities in the same event.

The close encounters of Evelyn (described earlier) are examples of this overlay of the physical and the metaphysical. They contain many elements to indicate that they are solidly rooted in nuts-and-bolts reality. Yet they also have a "dreamlike" metaphysical quality to them, the surreal sense of unreality. It is this paradox, the transformation of the reality of the experiencer that most defines the event of the fourth kind.

"Where are we?"

"What's going on?"

I often hear these questions during hypnotic regression, as the experiencer relives the events leading up to a UFO abduction. During the moments surrounding an encounter of the fourth kind, many experiencers describe this sense of alteration to their normal reality. Their world becomes surreal. Their reality begins to change and they begin to enter the realm of the close encounter.

Most often, this transformation seems to occur during the beginning of the encounter. It might be perceived as a small anomaly, in which some minor aspect of the environment doesn't make sense. It might be a larger scale event, in which an experiencer notes that the overall environment has somehow changed - the surroundings suddenly becoming quiet or unreal. Or it might be a full-scale reality transformation, in which the experiencer is entirely removed from the everyday world.

Small anomalies are those in which a relatively minor element of the environment "doesn't make sense." While most of the world is normal, some minor aspect of it behaves in a paradoxical or illogical way. In one case, a witness described to me how, as she and a friend drove across the rural countryside late at night, a "red traffic light" followed their car for many miles. The event began as their car initially accelerated from a traffic light early into their journey. After some distance, they noticed that the light was still behind them. It was observed by both witnesses for many miles. The light seemed to "follow" the car through a number of turns.

In another such case Evelyn, upon her return from her "Pulled out of Bed" CE4 event (described in chapter 2), remembered watching as lights departed from above her house. She observed them through her kitchen window, which she clearly remembered was opened outward. On subsequent inspection of these windows, it was observed that it was not possible for the window to open in the manner in which she remembered.

In both cases, we can think of each of such relatively minor anomalies as a superposition of realities, the reality of the encounter, and the reality of our everyday world.

Also common in our case files are somewhat larger anomalies, in which the overall reality of the experiencer appears to be altered, but the experiencer has apparently not (yet) been removed from the environment. Often the experiencer will describe feeling a sense that "the world is wrong..."

The most commonly described of these phenomena is what Jenny Randles termed "The Oz Factor."[xv] Often on the entry phase of an encounter, the environment seems to take on an eerie, silent, "Twilight Zone" character. Sounds in the environment seem to be absent; none of the expected frogs, crickets, wind in the trees, etc., are present. Evelyn, Janice and many other experiencers have all described the same eerie silence, which seems to precede a close encounter.

Selective Viewability

Another related type of reality anomaly is the effect of "Selective Viewability" which we noted in chapter 1. In such cases the experiencer is able to very clearly perceive the UFO. However, others who are nearby apparently are not aware of it. There are many cases of this type in the literature, including several on the Minnesota MUFON website[7] and the Triangle Sighting described in chapter 1.

Characteristic of these is the fact that the witnesses who do perceive the object generally give similar descriptions to an investigator, suggesting that they clearly observed the same object. The object is often described as "obvious" to those who perceive it. In addition, those who don't perceive it are often in a situation where "they couldn't possibly miss it" if the object were actually visible to them.

---

[7] The Minnesota MUFON website is located at: http://www.mnmufon.org

These anomalies pose the question of what allows one person to observe a phenomenon while others in nearly identical circumstances do not. Are these simply examples of the limits of human observation under unusual circumstances? Or are these clues to a deeper metaphysical aspect of the UFO/CE4 phenomenon, possibly of reality itself, which we do not yet understand?

Reality Transformation

The third and most profound type of anomaly is that of the full reality transformation, the key element in any event of the fourth kind.

Typical of these is one case that is described earlier, in which an entity appeared in a ball of light to Evelyn, who was lying in bed. She was then drawn into the light. While there, she observed that the interior of the ball of light was considerably greater than what could be accommodated by a room of that size, implying that some form of "transformation of worlds" had occurred.

Several cases have been described to me in which a witness was driving on a major highway in Minnesota. In each case, the witness suddenly found him/herself on a stretch of road, which appeared to be "different" or somehow "wrong." In one such case, thinking that he and his passenger had somehow become lost, the witness exited the freeway and decided to try to figure out where they were. As they drove up the exit ramp, he felt that, in his perception, the world looked somewhat like a stage set that had been only partially completed. This sense of unreality was then followed by an interval of missing time, after which, he found himself back on the correct highway.

Was this the entry scenario of a CE4? The events in this case strongly suggest that it was indeed the beginning of an abduction event.

Janice's "Wolves on the House" encounter (described in chapter 2) is another powerful example of a reality transformation. In one instant, she was standing on her lawn, observing strange beings - apparent "wolves" - sitting atop her house. The next moment she found herself in the alien realm, the object of apparent scientific experiments.

Understanding Reality Anomalies

There are several possible ways we might be able to understand reality anomalies. One possible way, suggested by Budd Hopkins and David Jacobs, is to ask whether such anomalous perceptions might actually be screen memories for another event. Most memories of reality anomalies are remembered under conscious recall and are not hypnotic. Thus, partially blocked memories of an event could result in a distorted perception of the experiencer's reality.

While the literal accuracy of memories from hypnosis cannot be assumed, hypnotic regression can often resolve many reality paradoxes.[xvi] Often these turn out to be entry scenarios into a classic abduction. While this is a current area of research, much work by Budd Hopkins (Intruders), David Jacobs (The Threat) and others, suggests that this is the case.

However, another view is that these reality anomalies are not an illusion. This view instead suggests that they are indications that a physical/mental duality actually exists. Based upon Jungian psychology,[xvii] this is somewhat the view taken by Jacques Vallee[xviii] and others, that UFOs and similar paranormal phenomena comprise a duality of physical and metaphysical reality. This view hints at a deeper underlying relationship between consciousness and the underlying reality of our world.

Whether real, as suggested by Jacques Vallee, or illusion, as Hopkins and Jacobs theorize, reality anomalies are truly signs of a deeper mystery. If we can crack this mystery, it might well lead us to a vastly deeper understanding of both the UFO/CE4 phenomenon and perhaps of reality itself.

Spontaneous Invisibility

Another strange effect noted by investigators of the close encounter phenomenon is that in which the experiencer finds him/herself seeming to shift out of reality itself, becoming invisible to their fellow humans. Several researchers, among them, Budd Hopkins[xix] and Donna Higbee[xx], have noted this effect and dubbed it "Spontaneous Invisibility."

After reading through several articles on the worldwide web regarding this and listening to Budd Hopkins describe several cases of this nature, my interest in this as a potentially distinct phenomenon was enhanced. I had interviewed a number of close encounter witnesses who related accounts similar to those described by Hopkins and Higbee. Thus, my next step was to review some of these and begin to catalog them.

The South Minneapolis Blackout Sighting

One fascinating case (briefly mentioned in chapter 1) involves a UFO sighting, which occurred in south Minneapolis during the 1990s. During a power failure, the witness and several other people had stepped into the hallway. The witness then went to the front window, about twenty feet from where they had been standing, to look out onto the street. To her surprise, when she did she observed a large UFO hovering across the nearby building block. Startled for a moment, she turned to the others to point it out to them. However, she found that she was unable to get their attention. This state of affairs continued for several moments, during which they appeared to be completely unaware of her presence. She then turned to the window and saw that the UFO had vanished.

The events as described by the witness suggest that this case might involve missing time, although it may also simply be a momentary, up-close sighting of a very spectacular UFO. However, at least one of the other people present indicates that, although they were aware that she had gone to the window, at that moment they were unaware of her presence there. Neither had the slightest idea that she was trying to attract their attention.

Additionally, this sighting occurred on a warm evening. Due to the power outage, a large number of people were outside on the street. These people should have been able to see the UFO. However, apparently no one except the witness (and possibly one or two others) observed it (Note: this is one of several "selective viewability" cases, described in chapter 1). Thus, if we are to take these witness descriptions at face value we are faced with dual phenomenon: Selective invisibility by the phenomenon and corresponding invisibility of the witness who, by some means, seems to be able to penetrate the UFO's "cloak of invisibility."

The "Tennis Court" CE4

Another fascinating case involves an abductee in western Minnesota. This person was outdoors, playing tennis with several friends of similar age. Several of them had taken a break from the game when the witness looked up to see a large silver disk hovering over a nearby building. Startled, he called out to his friends to bring the object to their attention. However, they did not respond to his shouts. He later told me that he was just about "in their faces" attempting to get their attention but that they seemed simply unaware of his existence at that moment.

At that moment, it appears that a period missing time began as the witness underwent a close encounter of the fourth kind. It is not clear whether the episode of spontaneous invisibility occurred before or after the abduction, or perhaps both. Whichever is true, it is clear that all of the people on the field ought to have clearly seen both the object and the witness. However, neither was visible to them during the time of the encounter.

The Invisible Driver

In yet another fascinating case of the relationship between the close encounter phenomenon and that of spontaneous invisibility, an experiencer, whom I will refer to as "Amy," has described to me a chain of events, which began with a tremendous close encounter and continued with a powerful sequence of psychic episodes. This time in her life was characterized by a sudden and profound parapsychological awakening, what she describes as a powerful, life-altering shift in consciousness.

One day, at the time when these changes were at their peak, Amy found herself in an experience while driving her car. She was stopped at a traffic light at a busy intersection. Her car was approximately third in line at the red light. In addition, there were a number of pedestrians on the nearby sidewalk. She suddenly noticed that several of the pedestrians were staring curiously into her car window. She heard exclamations such as; "There's no one in that car!" and ..".Who's driving?" This continued for several more seconds, until the light changed and she accelerated through and past the intersection. This was just one of several incidents that occurred to her during that time frame. Similar to other events described in this chapter, her invisibility experiences were somehow associated with a period of close encounters and of tremendously heightened PSI experiences.

What are these anomalies in which the experiencer seems to vanish temporarily from our reality? Like the reality anomalies described in the previous section, these experiences seem to imply that there is a key aspect of our reality that we do not understand. And within this riddle may be another key to the strange enigma that is the close encounter.

Learning from High Strangeness Cases

As we stated earlier, the best way to understand an aspect of our world is often to study it in its extreme cases, where it seems to work in unusual or problematic ways. Thus, perhaps ultra-high-strangeness events such as reality shifts and spontaneous invisibility offer us a glimpse into the deeper nature of reality. We can learn how it is affected by the UFO phenomenon and perhaps learn more about its interactions with human consciousness.

It is my belief that ultimately, this will teach us much both about UFOs and of reality itself - and of the role the experiencer plays in bridging the normal and the extraordinary realms.

Problems for the "Nuts and Bolts" View

We have discussed the nature of the phenomenon to some degree but have so far avoided the question of who/what the visitors actually are. Some of the comments in this book and of UFO researchers in general, imply that the mechanism behind the phenomenon is one of nuts-and-bolts visitors coming here in physical spacecraft, etc. Other encounters suggest that the events are non-physical or paranormal.

So, which is it? Do the apparent actions of the visitors imply any specific origin or nature? As we have seen in this chapter, the phenomenon presents us with a host of paradoxes. Let's examine some of these paradoxes and see what these imply. Is the phenomenon a physical one, or is a deeper understanding required?

Where are they from?

During their encounters, many abductees have asked the visitors where they come from. The responses have been many and varied. During hypnosis sessions I have been told by experiencers that the aliens come from "Zeta Reticuli," from "Andromeda," from "another dimension" and a host of other origins. Most researchers have described similar results. On multiple occasions, I have noted Budd Hopkins and others to state simply "Aliens lie." What does this mean? Is there a deliberate alien attempt to confuse or otherwise obfuscate regarding this question?

From experiencers' descriptions of their encounters, there seems to be a "trance logic" involved. It is not the logic of the detective or the scientist but seems to be more characteristic of a dream-like state of consciousness. It is the logic of metaphor and symbolism. "Facts" which come from this state of consciousness often do not appear to follow the rational laws of mathematics or logic. They are, in a sense, "dreamlike." Yet these dreams can sometimes leave behind physical evidence, suggesting that they are fully real.

What does this imply? Does this somehow reduce the hard reality of the visitors? Or does it suggest that there is something about the visitors we simply do not understand? My suggestion is that the latter case is true. To me this further suggests just how little we understand about the close encounter phenomenon. I often wonder whether a better understanding of this apparent quasi-information related by the visitors, themselves might hold another important clue to understanding the close encounter phenomenon. We will re-examine this apparent non-rationality of the visitors later, when we discuss different models of visitor contact.

The Air Traffic Control Problem

Several contemporary researchers have suggested that the UFO phenomenon is a purely "nuts and bolts" phenomenon, using technology which, although advanced, is understandable in principle.

In chapter 2, we discussed the apparent scope of the close encounter phenomenon and its interaction with humanity. We observed that UFO abduction seems to be extensive. Some researchers have even suggested that this apparent "program" of UFO abduction has been steadily expanding throughout recorded history, especially in recent times[xxi]. Indeed, many experiencers have suggested this to me during interviews and/or hypnotic regressions.

This massive scale presents an issue that any model of a physical presence of the visitors must address. It is what I call the "Air Traffic Control" problem. Simply put, if abduction is a purely physical phenomenon, then a tremendous number of alien craft must be required to sustain such a high volume of abduction. Thus, the phenomenon must be truly massive in scale.

Let us accept for the moment, the premise that the visitors are purely physical in nature. From abduction reports alone, we can estimate the scope of this phenomenon and see how vast it is.

According to several different sources, such as the Roper

polls of unexplained experiences[xxii], up to one percent of humanity may be abductees. One researcher claims that for each abductee there are about two hundred abductions in a lifetime (!). These events occur over an average life span (which we will assume to be about 75 years), beginning almost at birth. I will conservatively estimate about 150 events occur during this time, resulting in approximately two experiences per year. From these figures alone, we can see the apparent scale of the phenomenon.

One percent of the population, each experiencing two events per year, gives us two events per hundred people per year. Dividing this number by 365 days/year yields approximately one abduction per twenty thousand people per 24-hour period. A large metropolitan area (such as Minneapolis/St. Paul) might contain on the order of 2 million people. With this rate of abductions, we find that two million people, times one event per 20 thousand people per day, results in about 100 abductions per 24-hour period in the metropolitan area.

Let's arbitrarily assume that one UFO flight is required to perform 10 abductions. This could reduce the number of UFO flights to 10 per night over a metro area. However, it still implies that the urban skies should be very busy with UFOs.

Let's further assume that this abduction rate of one per 20,000 people per night applies worldwide. We know that there are about 6 billion people in the world. We can therefore calculate that there are 6 billion people, times one abduction per 20,000 persons per night, which results in 300 thousand abductions per night. If we again assume about 10 abductions per UFO flight, we would need 30 thousand UFO flights per night, every night, to sustain such a global UFO abduction program.

What the Numbers Tell Us

The numbers described above, although rough estimates, clearly reveal what I refer to as the "Air Traffic Control Problem." They reveal a phenomenon that, if entirely physical in nature, must be truly massive in scale. In order to coordinate such a massive abduction program and even to prevent UFOs from running into each other, a significant ET infrastructure, including a UFO air-traffic-control system, must be required. Unless this phenomenon truly is as massive as we have estimated, something about our understanding of UFO abduction and of our paradigm of UFOs and abduction as physical events, must be incorrect.

Based upon these numbers, we can draw some further conclusions about such a nuts-and-bolts abduction paradigm. The first thing we note that there would have to be many thousands of UFOs in the air, on the ground, or in near-Earth space at any given time. To estimate just what kind of a support effort this would require, let's make the (very human) assumption that these UFOs would also need some sort of ground crew to perform maintenance. If we assume a support crew of one to ten ETs per UFO, we end up with tens to hundreds of thousands of ETs. This would require a huge base or mother ship, which would presumably be massive and difficult to conceal.

The numbers imply that unless this vast infrastructure exists to support the rate of UFO abductions, one or more of the following must be true:

1) Each abductee has one to two orders of magnitude fewer experiences than the highest estimates claim.
2) The number of abductees is far less than the suggested one percent.
3) UFOs are able to escape detection except on very rare occasions. This would imply that they had some visual and radar evasion technology, perhaps a reasonable speculation if we extrapolate our own stealth technology a century or two into the future.
4) Large numbers of abductions appear to be nonphysical or metaphysical in nature. Thus, they may not require a physical UFO visit. This would argue (at least in part) against a physical nuts-and-bolts nature of UFO abduction and would imply that it is in some way less material.

It is not clear which of these possible resolutions to the "Air Traffic Control" problem is the correct one. Perhaps all of them are to some degree valid. My own view is that an understanding of the resolution to this question is one of the "master keys" to our understanding of the UFO/CE4 phenomenon.

Ambiguity - The Standard Model and More

The frequency of non-standard encounter events; the ambiguous origin of the visitors and the "Air Traffic Control" problem all suggest to me that the UFO and close encounter phenomenon must be far more complicated and far more ambiguous than we think. Yet these issues are only a sample of the strangeness that is the close encounter phenomenon.

We have seen that there is a core reality to the phenomenon, a set of common themes, which we called the "Standard Model." Yet, in this chapter, we have seen cases with variations from the standard model.

Are the standard model and other models mutually exclusive? Or could more than one paradigm be true at the same time? The ambiguity and the wide variations of experiences described suggest that at its core the phenomenon does not have a simple explanation.

Several researchers now claim that we are nearing a point where we can find the answer to the CE4 mystery. They paint a very well defined portrait of the close encounter phenomenon - what it is, who the entities are and why they are here. But is there one core truth to it? Or is the phenomenon more complex than we have even yet imagined? I feel we are just beginning to address these questions.

To me, the high degree of strangeness of many cases implies that, regardless of the origin of the visitors, the nature of the phenomenon is something extraordinary. Thus, I believe that we cannot assume any form of conventionality of nature and/or origin of the phenomenon.

For all of our years studying the UFO phenomenon, I suspect that we still know very little about the reality of close encounters. But from examining the phenomenon in extreme cases, we learn that it has endless surprises in store for us.

In the next chapter, we will look at the effects of the close encounter on the experiencer - the human side of contact. We will begin to examine more deeply, the relationship of people with the phenomenon. In subsequent chapters, we will examine this relationship in more detail.

Extraordinary events such as those we have examined in this chapter stretch our understanding of our world. They blur the boundaries of our worldview and challenge the reality of the experiencer. Thus, for the experiencer, they form a bridge between the everyday world and that of the extraordinary. It is a Cosmic Bridge to the other side of the sky.

Chapter 4 – The Human Side of Contact

The close encounter can be either a terrifying or a wondrous event. In the last two chapters, we saw how profoundly the phenomenon can affect the life of the experiencer. For many, it is a curse called "abduction." It is a time when one is confronted, overpowered and taken against one's will by unexplainable, unearthly beings for reasons not understood - a dark and sinister force.

Yet for others it is an honor, a moment of wonder at contact with a power greater than ourselves. While frightening, it can be a time of communion and perhaps even of enlightenment, a chance to meet face to face with a source of wonder in our universe.

Especially in encounters such as those experienced by Evelyn and Janice, such events initially resulted in mystery and fear. Yet in each case, a legacy of growth and awakening has resulted as each experiencer has come to terms with these life-transforming events.

For better or for worse, the individual close encounter seems to be a focal point of the interaction between humanity and the phenomenon. Humanity is experiencing contact, one individual at a time. So in this chapter, let's examine some of the effects of the phenomenon on the life of the experiencer. In so doing we can set the stage for later examination of how the phenomenon interacts with humanity as a whole.

Contact: The Experience and its Impact

I am repeatedly struck both by the effect that the close encounter phenomenon has on peoples' lives and by its pervasiveness in our world. Early in my studies, as I began to meet more and more close-encounter experiencers, the impact it has had in their lives quickly became apparent. To all who are "chosen" for contact, the phenomenon can be life altering, or even shattering.

Some see the close encounter phenomenon as dark and some as light. Yet, in most cases, the events described in encounters are quite similar. So what makes the difference? Could the emotional nature of the experience be affected by our own consciousness? Do our own perceptual filters affect the apparent good/evilness of the phenomenon? How does the phenomenon relate to the growth of the experiencer and the sense of psychic and/or spiritual awakening that so many seem to experience because of their encounters?

I often give presentations on UFO encounters at civic groups, science fiction conventions and new age expositions. Recently, I conducted a workshop at a local science fiction convention in Minneapolis. As the talk progressed, I covered UFO-related topics in increasing levels of strangeness. I noted that as I moved into the emotional impact of close encounters, the attention of several people became increasingly riveted on the front of the room.

As I began to describe the effects of the phenomenon on peoples' lives, I noted that one person in particular suddenly had tears running down her face. After the workshop, this person and I talked for several more hours. During this time, she described to me a history of unexplained experiences, strange effects and puzzling memories that had dogged her all of her life. She told me details of the many ways in which the phenomenon had affected her. She described how uninvited and inescapable these intrusions were.

You are not alone

In my conversations with close encounter experiencers, as a UFO researcher and as simply as a sympathetic ear, one of the most common complaints I hear is how experiencers feel isolated from those around them. They lack someone they can tell about their experiences without fear of ridicule. They fear that others will conclude they are crazy. But exactly how alone is the experiencer? A quick look at some numbers suggests that the phenomenon is very extensive. There are far more encounters and many more experiencers than one might think.

While we can only speculate on the exact numbers, we can make some rough, order-of-magnitude estimates. In so doing we find numbers that are surprising. The Roper Polls of Unexplained Experiences[xxiii] have estimated that from two-tenths percent to one percent of the population have had enough unexplained events in their life for us to consider the person to be an experiencer. In a metropolitan area of two million people, this would imply that somewhere between four thousand and twenty thousand people might be close encounter experiencers, a staggering number to consider.

Thus, we can no longer think of such encounters as rare, isolated events. Instead, they are common within our society. To those who are not experiencers, this may be startling. For those who are, and who feel isolated, it may be a comfort to know that there may be more than you think. You are not alone.

Research and Understanding

One of the first issues to arise after having had a UFO encounter is the experiencer's level of comfort with disclosing that fact. Who can one trust? Who will listen to the account without judgment or ridicule? Often the first step is to tell a trusted friend, sibling or other personal confidant, someone who will accept the experiencer's account of events.

If one feels comfortable talking with a researcher, there are a number of options and the UFO research community will welcome the report [See appendix A for a list of UFO research agencies]. In cases of contact, it is often customary to conduct research confidentially, where the account is shared only between the experiencer and the researcher. This allows the experiencer to better control the degree of disclosure of their account, yet also helps the researcher to gain an overall picture of the phenomenon. The most important criterion is the experiencer's level of comfort in discussing these events.

For many experiencers however, merely reporting a UFO encounter is not enough. As we saw in the chapter 2, there may be a legacy of fear. The event may have been traumatic and its effects long lasting. In such cases, it may be important to seek some form of assistance. The investigator may be able to suggest where the experiencer can find assistance. [Note: Appendix A lists organizations, which may be of assistance in coming to terms with close encounter experiences.]

61

The Fear Factor

Many experiencers describe the UFO phenomenon as a source of fear in their lives. They dread the next appearance of the visitors and for some, even subtle reminders of encounters result in fright and alarm. What is the nature of this fear factor and how can one come to terms with it? When is it a natural reaction and when does it become a burden in the life of the abductee?

The most important step in dealing with fear is to understand it in perspective. Fear is a natural response when confronted with the unknown and there are few events more "unknown" than an encounter with alien beings. Thus, perhaps a certain amount of fear is to be expected.

However, for some experiencers, fear can show up in situations unrelated to the encounter(s) but which might suggest aspects of past UFO events. Often, such fear may be far greater than might be warranted by the situation.

Examples might include the unexpected passage of a low flying aircraft, seeing the planet Venus or another bright star under just the right (wrong) conditions, noticing a book on UFOs with a picture of an alien on the cover, etc. This heightened fear factor is referred to by some authors[xxiv] as "Post-Encounter" or "Post-Contact" Syndrome. In some cases it can result in sudden, disproportionate fear responses. In such cases, fear management becomes top priority in the healing process.

Each of us has our own perceptual filters and in some cases, negative post-contact effects may be the result of the subconscious ways in which the mind processes the events of an experience. Thus, an important step in coming to terms with extraordinary fear is to understand the source of a person's perceptual framework. Is the fear due to the encounter itself, or is it due to other factors in the experiencer's life? Events such as past trauma (which may or may not be related to UFO encounters), or early childhood learning can create subconscious influences that could enhance the fear response.

Evelyn, Taming the Fear

Can we, using a healing technique such as hypnotherapy, help to reduce the fear factor in close encounter experiences? In the case of Evelyn, described earlier, she had developed a number of fear reactions subsequent to her dramatic series of close encounters. These included her spectacular, but only partially remembered, "pulled-out-of-bed" scenario (described in chapter 2).

Evelyn filed a UFO sighting report with MUFON. She was interviewed by MUFON field investigators, providing a large amount of valuable data on these events. During these interviews, an extensive chronology of UFO-related events emerged, along with a deep sense of fear. She described feeling frightened by things that reminded her of her encounters. As she described her experiences, her fear seemed to grow until it seemed an almost physical presence in the room. Its effects on her life were causing her difficulty. She found herself becoming anxious when she was alone in the house, or in situations that reminded her of the UFO topic.

Our initial field investigation had occurred before I began my study of hypnosis. Some time later, when I became certified as a hypnotherapist, we arranged a session. In this particular case, we conducted a hypnotic regression in order to help her better understand these events. However, each time her mind approached the events, she found herself paralyzed by fear. As we began the first hypnotic work, the fear again showed itself with its full brutal force. I could only admire her courage as she, despite a look of terror on her face, told me she still wanted to continue with her regression. Evelyn wanted to get to the bottom of her experiences, to put them to rest once and for all. Thus, we decided that in the next session we would focus on managing the fear she felt. Only then would we continue with regression work.

During this follow-up hypnotic work, other factors, not directly related to her experiences, emerged as contributors to her fear reaction. A part of her mind, associated with normal childhood memories and religious beliefs, perceived the close encounter phenomenon as evil, a thing to be feared.

With Evelyn in deep trance, we were able to use hypnotherapy to help reduce this fear factor. This permitted her to get past this obstacle and continue with the regression. As a result, she was able to uncover and explore the long and meaningful series of close encounters that had permeated her life.

As with any aspect of life, there were both positive and negative elements to Evelyn's encounters. Yet from that point on, the phenomenon would be more objective and manageable to her, while remaining as a deeply meaningful part of her life. The new understanding that emerged from these regressions and her greatly reduced levels of fear, brought her a renewed confidence that has continued to grow to this day.

This case illustrates how one of the most important factors in resolving strong fear reactions is to find and resolve the deep source of the fear in the experiencer's mind, influences which may or may not have resulted from their actual encounter experiences.

A Frightening Night Drive

UFO encounters can happen nearly any time or place. Some may be abductions, while others are simply sightings. Yet each is an unexpected intrusion of the unknown into the life of the witness. We have already seen how the sudden appearance of the unknown can prompt fear. For some, this fear can be extreme.

A number of such cases exist in the Minnesota MUFON case files. In one representative case, two motorists named "Mark" and "Terry" (pseudonyms)[8]. The two were driving through the countryside some time after sunset, heading home after a day in a nearby city. Mark suddenly pointed out a distant light in the sky that seemed to hover far ahead of their car. As they watched, the light slowly descended toward the highway. Then a turn in the road hid it from their view.

When they next saw the object, it was huge and it was close, hovering just above the tree line to the right of the highway. Both of them were frightened. Terry, who was driving, floored the gas pedal and they sped off down the highway toward home. As they did, Mark observed that the object appeared to be following them, always just over the tree line to their right and slowly descending. Eventually they reached a small town close to their home and the object was lost to view.

On investigation, it became apparent that object's path and description, as well as the geometry at each stage of the sighting, corresponded closely to the observation of a small aircraft.[9] In most sighting cases, this might have been the end of it.[10] However, one interesting aspect of this case was the fear that the witnesses felt as they observed the object. As we indicated earlier in this chapter, such fear reactions can suggest possible past encounters. I

---

[8] This case actually is an aggregate of several similar cases. The details of these cases are obscured to maintain confidentiality of those involved.

[9] The exact details of these cases are withheld to preserve witness confidentiality.
[10] Note: As with most UFO sighting cases, one can never say with absolute certainty what had occurred, but this was our best hypothesis.

therefore wondered if, in their lives, there might have been more interaction with the phenomenon than they had indicated.

As I investigated this sighting, I also interviewed the witness's other family members. During these interviews, several of the family described multiple UFO encounters, including close UFO sightings and events that suggested possible close encounters. Several family members described events similar to those described in the chapter 2, suggesting UFO abduction activity. Thus, it appeared that this case might involve far more than just a single misidentified object.

One question investigators routinely ask witnesses is how much they had read or heard about the phenomenon. In this case, the family had had little exposure to information about UFOs, but had heard much from a minister who taught that UFOs were demonic. Could this exposure, both to the phenomenon itself and to admonitions that it was evil, have combined to enhance the fear response of these two witnesses on their frightening night drive?

Positive on the Surface, Dark Beneath

Another group of nearly identical experiencer cases can be represented by the story of someone I will call "Jim," who has had unexplained experiences since childhood.[11] As far back as his childhood memories go, a playful cartoon-character companion had visited him. This friend would appear to him at irregular intervals, playing and telling him stories and taking him away on adventures. Jim often found that his little friend could be a welcome refuge from a challenging world.

As Jim grew up his companion seemed to disappear, to be replaced by other events that left him with a deepening sense of mystery. However, his overall view of his encounters was generally positive. He enjoyed life and said that he also enjoyed his interaction with the phenomenon.

---

[11] The case of "Jim" is actually an aggregation of several nearly identical experiencer accounts.

I have noted that sometimes, just as when excess fear is involved, an overly optimistic view of the phenomenon may suggest that all is not as it seems. Jim's experiences were no exception. At several points, as he described his "fun" experiences to me, he broke into tears - not tears of joy. Was there a previously hidden dark side to his experiences?

At Jim's request, we began hypnotic sessions. In deep trance, he began to examine his "fun friend." He quickly noted that his friend had large black eyes. He continued on, describing his companion as a small two-legged creature, about four feet tall, with no hair and a big head, a classic "small gray" alien (see chapter 2). His fun and companionship with this character had become a classic close encounter.

As we continued, Jim recounted a lifetime of abductions, very similar to those which Evelyn and many other experiencers have described, and which permeate the UFO literature. His description of them initially had a very dark, frightening tone. However, as he continued to describe his encounters he began to examine the worldview through which he saw them. As he did, he increasingly spoke in terms that were more objective. He found that his polarity of feelings, the joy of his "companion" versus the darker tone of his encounters, was steadily changing into an increased understanding and resolution. He found himself losing the surface innocence but healing the pain beneath. The phenomenon was increasingly beginning to fit into his life in a more manageable way.

Intent: Good, Evil and Awakening

Among the variety of encounter events, some are described as dark, ominous intrusions, while others are described as more benevolent or even enlightened. This duality of good and evil seems to be a strong theme in the UFO phenomenon, as if they were two sides of a coin. On one side is the idea that the visitors are genetically exploiting us. David Jacobs takes the dark side concept the farthest in his book, *The Threat*, in which he portrays the phenomenon as a means by which the aliens plan to take over the world.

On the other side of the coin is the idea that the purpose of the phenomenon is to provide nurturing, guidance and assistance to humanity. In this view, the focus is on the spiritual growth of the experiencer and of humanity as a whole. Some experiencers describe how, for unknown reasons, the visitors feel it is important that we mature[12] at a faster rate. This may be to prevent us from destroying our own world. But perhaps it might also be to bring us to a point where we are no longer a threat and can be welcomed into some greater community "out there." This view is taken by much of the "positive" community and is related in books such as *Preparing for Contact* by Lyssa Royal and Keith Priest and *Healing Shattered Reality* by Alice Bryant and Linda Seebach.

Most commonly however, experiencers describe their encounters as a "mixed bag." Both positive and negative events seem to occur, frequently during the same encounter.

---

[12] Maturity, in this case, appears to mean spiritual enlightenment and/or ethical conduct of our affairs. We will discuss this in more detail in future chapters.

One event which initially appeared to be a classic abduction, turned out to have many "surprises." In this event, the experiencer, whom I will call "Nancy," recounted to me under hypnosis an event in which she was abducted in the typical manner (see chapter 2). Instead of being subjected to the "standard" experience, she was taken to a large craft, where she was given a lesson on the environment and future apocalyptic events. At the conclusion of the scenario, Nancy noted that the place where she had been taken was "an education center." What was the purpose behind this "lesson?" They appeared to be concerned for our welfare. But were these entities working for our benefit or otherwise?

While deep in trance, another experiencer I will call "Anne" described a scenario in which she underwent a medical procedure at the hands of "spiritual scientists." She was abducted and taken to a UFO. She was then seated in a room in which the scientists lectured her on some type of advanced mathematics.[13] The entities appeared to be humanoid, wore long robes and had facial features that were partially, but not completely human. At approximately the same time, two smaller entities behind her conducted a medical procedure on the back of her head, which they referred to as "giving her brain a checkup."

During this same regression, as we traveled back through a chain of earlier similar experiences, Anne also found herself reliving an experience in a metaphysical domain, in which she was a non-human entity. She felt that this had been in an earlier lifetime but was somehow associated with her current experiences. It emerged from the regression that the phenomenon's purpose in her life was largely spiritual, associated with her learning lessons and fulfilling a role as teacher and healer.[14] The present-day abductions were but a part of this personal growth process.

---

[13] Another therapist, upon seeing samples of the writing which Anne reproduced, indicated that it looked like astrological symbols, rather than mathematics.
[14] Anne works in the medical field and is also a certified hypnotherapist.

While in both of these cases aspects of their experiences were unpleasant, their relationships to the phenomenon appear to have a longer-term positive tone. Both experiencers have described many similar (but not identical) aspects to their encounters. In addition, each has described how, in the long term, their encounters have profoundly deepened their spiritual lives. While many events were frightening, they consider their overall encounters to be positive events. Thus, rather than being good or evil, experiences such as those of Nancy, Anne and others seem to vary, or even to defy the good/evil polarity altogether.

Perceptions: Resolving the Polarity

In these cases and in many more that fill my files and the files of other researchers; we see how one's worldview can strongly affect the experience (both positive and negative) of the phenomenon. If these factors are extreme in one way or another, the experiencer can feel caught between positive and negative sides of the phenomenon - hope/belief that contact is positive in nature versus the fear and trauma of the dark side of abduction. However, as one better understands one's own perceptual filters, this duality between light and dark side can resolve and one can deal with the phenomenon more objectively.

What is the ultimate motivation behind this anomalous presence among us? While this remains strangely ambiguous, there is a consistent theme and suggestions of a coherent purpose to the phenomenon - a reason for which it interferes in the life path of the experiencer. In later chapters of this book, we will examine these possible motivations and suggest some possible strategies that may be at the heart of the visitor interactions with humanity.

It is not clear whether the visitors mean us good or ill. Yet in this chapter, we have seen that close encounters can be either positive or negative, or sometimes both - a combination of trauma and growth.[xxv]

In this chapter, we have examined the human side of UFO encounters, the individual experience of visitation. We have seen the wonder and trauma of contact. We have also noted how, for many experiencers, the task has been to resolve these two sides of the experience.

Each may experience contact in a different way. Thus, each experiencer must come into relationship with the visitors in their own way. For each, it is the building of their personal Cosmic Bridge.

Chapter 5 – The Contact Relationship

Who are the visitors, and what is their relationship with us? This may be the fundamental question of UFO studies. Is the phenomenon good? Is it evil? Or is it something different entirely, beyond the bounds of human experience? Might it even be something that the human mind is unequipped to grasp?

In the first few chapters, we introduced the idea that we can think of the UFO phenomenon as a relationship between the witness/experiencer and the visitors. In later chapters, we discuss some possible agendas of contact and speculate on the ultimate goal of the visitors. In this chapter, we explore in more detail, the idea of the contact relationship - both as described in the literature and from the point of view of the individual experiencer.

We ask whether a possible "agenda of contact" might be to build some form of a relationship with humanity. If so what would be the goal of such a relationship? As an individual, what does it mean to be in contact, either positive or otherwise, with the visitors? What might this imply for the overall relationship between humanity and "them?"

The Contact Process

In chapter 4, we discussed the human side of the close encounter phenomenon and observed that such contact seems to occur throughout the lifetime of the experiencer. It appears to be a continuous, lifelong process, similar to education, career or social relationships. So perhaps our first step in understanding contact is to understand this long-term, ever-evolving relationship between the experiencer and the phenomenon.

The close encounter phenomenon is about people. Nearly all close encounter researchers believe that it focuses on the individual experiencer and follows him/her throughout life. Nearly every experiencer has described how the phenomenon has manifested in their life in ways that evolve over their lifetime. As the experiencer travels through life, from childhood through adulthood, their interactions with the phenomenon assume corresponding phases as well.

- Childhood

During childhood, the phenomenon seems to take on a playful air, as if the visitors were presenting themselves as playmates as well as teachers. The emphasis appears to be on developing, teaching and testing the child. This might be with special lessons during abductions, by exposing the child to psychic toys, or by the visitors posing as "magical playmates."

Numerous experiencers, such as "Jim," who we met in chapter 4, have described how they first encountered the visitors as a child, when one or more entities would appear as playful companions.

Also in childhood, the experiencer frequently begins to feel a sense of separateness from other people, to feel that they are special. Evelyn, Janice and nearly every other experiencer have told me of the sense that, from childhood, they were in some way different from their peers. Some abductees say that they feel their true home and family are "out there." Many abductees, including

Evelyn and Carolyn, have told me this. Both have described these feelings as a result of what they have learned during their experiences in the alien realm.

- Adolescence:
Some abductees describe how, during adolescence and into young adulthood, the phenomenon tends to take on a more serious aspect and more advanced learning begins. Often, experiencers draw an analogy to being in school or learning lessons.
    Though in my own research, I have found few cases of adolescent experiences, Whitley Strieber extensively describes such events in his book, *The Secret School.*[xxvi] In this book, he describes childhood experiences in which he and other children were grouped in what he called the "children's circle." In these encounters, a non-human teacher taught them about the environment, alien technology and related topics, lessons that would set the stage for his later encounters.

- Puberty into Adulthood:
At the time of puberty, the phenomenon takes on a more sexual or reproductive orientation. For a number of years, the focus shifts from the experiencer as a whole, to the experiencer's body. Often this is a time of trauma and at this point in their lives, many experiencers develop powerful emotional reactions to the phenomenon. Experiencers, such as Evelyn, Carolyn and others, all describe this aspect of the phenomenon, with its medical/reproductive focus and its corresponding impact. We discuss this in greater detail in chapter 6.

- Adult life:
    As close encounters continue throughout adult life, many experiencers describe how their relationship with the phenomenon begins to take on a deeper, more spiritual form. Encounters may have a more metaphysical nature. While not always the case, encounter events tend to assume a more positive tone.

As an experiencer becomes older, he/she often begins to remember encounters in increasing detail. While the effects of this can be emotionally powerful, in many cases we can see this as a positive influence. At this stage, many begin to seek out researchers and therapists to help them come to terms with the memories of their encounters. Evelyn and Carolyn were both adults, nearing midlife when I conducted much of my work with them. For each, it was time to shed some light on the mystery that had followed them for most of their lives.

Additionally, as many experiencers come to terms with the phenomenon, I have often observed them to shift their worldview, often toward an eastern or new age belief systems.[15] Many also become very concerned about environmental topics and believe that this is related to the mission they are here to perform.

Evelyn, Carolyn and others I have worked with have expressed this new age metaphysical perspective. Each has also described increasing concern about the state of the world environment and the ecological crises they believe lie ahead for humanity. Nearly all feel they have a special mission or knowledge, often with respect to the environment or nonspecific "Coming Changes." In the next section, we will further explore this awakening of a new perspective.

---

[15] This might also be due to my home and hypnotherapy practice being located in an area where such belief systems are more prevalent

Contact and Awakening

As we mentioned previously, many experiencers say that because of their encounters, their personal perspective has drastically changed and that they have become far more "spiritual" in outlook. Many also state that, they have developed increased psychic abilities. Researcher case files, as well as much of the literature, describe the phenomenon of psychic/spiritual awakening as being extremely common among UFO abductees.

Many abductees, who describe frightening experiences, also suggest that they are initiated along a psychic and/or spiritual path by their encounters. Yet the UFO phenomenon appears to be just one possible initiator of such a transition and many persons whose psychic talents have been awakened in other ways have described increased sightings of UFOs and other unexplained phenomena.

As opposed to abductees, contactees generally report experiences that are more positive. Often, they indicate that they have had other psychic experiences prior to their UFO encounters. They seem to be brought to their UFO-related encounters as a result of this path, rather than vice-versa. Thus the UFO experience appears to be one of many possible phenomena with a multi-faceted relationship to spiritual/psychic emergence – at once cause, effect and in parallel.

In our earlier discussion of Evelyn's encounters, we noted how she had been struggling for several years with the emotional effects of a series of close encounter experiences. We also noted how, in addition to the fear she experienced because of these events, she had also begun to experience intrusive paranormal effects in her life and that of her family. Watches and clocks would not work. Electronics around her would cease to function.

Several months later another experiencer, whom I will refer to as "Amy" (described in the previous chapter), also approached me to do hypnotherapy work with her. Amy was also dealing with an issue related to close encounters but apparently quite dissimilar to Evelyn's. While Evelyn was struggling with the fear engendered by the intrusion of the phenomenon into her private life, Amy was attempting to come to terms with a sudden, involuntary, unexpected psychic awakening.

While these two experiencers seemed to have very different relationships with the phenomenon, they had one thing in common. Both were attempting to come to terms with an apparent side effect of the close encounter phenomenon, which I refer to as "unexpected awakening." I have noted that nearly all experiencers describe this effect.

In the remainder of this section, let's examine some aspects of awakening and see how it is - and is not - related to the UFO and close encounter phenomenon. What can we learn from this effect? Is awakening a result of the close encounter phenomenon? Might these two phenomena occur together in some sort of lockstep relationship? Or is the psychic awakening aspect of the UFO phenomenon simply part of the greater psychic/spiritual emergence phenomenon often described in the literature?

What is an Awakening?

Nearly every abductee I have met describes this effect, which I will refer to as the "awakening" phenomenon.[xxvii] Cases of this sudden personal transformation fill my files as well as those of many other researchers. Just what seems to be happening to those experiencers who describe this phenomenon? What is a psychic/spiritual awakening?

Answers to this question will be quite different for different people. However, broadly speaking, the effects seem to fall into two categories. The first is the unexpected onset/enhancement of psychic abilities. Precognition, telepathy, clairvoyance and many other psychic experiences seem to be commonplace.

The second is a sudden change in spiritual and social outlook. Many experiencers I have interviewed describe how their viewpoints, especially on environmental topics, suddenly switched from being largely unconcerned to being very much concerned with the future of our world. At the same time, religious/spiritual views of many experiencers seem to become much deeper. Often they adopt an entirely new viewpoint, frequently eastern or mystical in nature. This new spiritual dimension in their lives seems to take on a very personal tone. Many, such as Amy and Evelyn, practice meditation or adopt a regular spiritual practice that is deeply heart-felt. Others such as Carolyn report become more empathetic to other peoples' feelings. Some describe, in the words of one abductee, "becoming closer to God."

I often observe experiencers associate the religious/spiritual change in their lives with the sudden emergence of their psychic abilities and frequently with the emerging memories of their encounters. The result can be a massive transformation of their social, professional, mental, spiritual and psychological foundations, their awakening.

Fear and Paranormal Fallout

When I first began working with Evelyn, I was struck by the depth of the impact that the close encounter phenomenon had had on her life. As described in earlier chapters, she had developed strong fear reactions resulting from these events. In addition, at about the time of her encounters, her outlook abruptly changed from that of an "average" person, who was relatively unconcerned with the outside world, to that of being quite interested in environmental, social and spiritual topics. She thus began to pursue these interests as she began to study for an advanced degree at a nearby college.

During this same time, she and her family also experienced a number of paranormal and metaphysical anomalies. Balls of light, UFO ground traces, sightings of anomalous creatures, electrical sensitivity, clocks that would randomly gain or lose time, numerous UFO sightings - the list went on...

As we commenced our hypnotherapy work and began to resolve the fear factor (described in chapter 4) we found that the phenomenon had apparently awakened within her the belief that there was a deeper mystical dimension to her life. Although dormant, this "deeper truth" seemed to have always been present in her life. As we explored her interactions with the phenomenon, it became apparent that it had been life long (she indicated that it had actually been over many lifetimes). However, it had been suppressed for much of her life and only now had she become consciously aware of this emerging facet of her life. She felt that she had interacted with alien and spiritual beings in many lifetimes and had a deep relationship with them.

Today, Evelyn has finished her teaching degree and is pursuing interests in metaphysical and historical studies. She is now largely comfortable with the memories of her encounters. She views her relationship with the phenomenon as life altering and, on the balance, very positive. For Evelyn, it appears that the fear-engendering abduction events might have been the "kick" that drove her to come to terms with what had in-fact been a long-standing relationship - her awakening.

Amy - Unexpected Psychic Gifts

Amy had felt overwhelmed at the time we first began our work. A few years earlier, she had experienced a powerful, dream-like experience, which appeared be a close encounter. In the days subsequent to this experience, she discovered that she had become very psychic, able to see/read the minds of the people around her.

It was as if she had been blind from birth, only to suddenly gain her vision late in life. Her life experience up to that point had not prepared her for this gift. Her talents were unpredictable and uncontrollable, a frightening influence in her life.

Initially, as we began therapy, the goal was somehow to control these manifestations. However, we quickly realized that the more we tried to "master" them the more resistance we got. Clearly, control and containment was not the answer. So we decided, "Why not just go with it?" Why not accept and embrace these psychic gifts?

I suggested to her a couple of books on psychic development[xxviii] with a number of excellent exercises to help people develop their psychic skills. In a follow-up conversation a few months later, she informed me that she was considering organizing a psychic development study group. By embracing her psychic gifts, she has been able to integrate them into her life in a meaningful way.[16]

Were Amy's newfound psychic abilities the result of a UFO abduction? Amy had few other indications to suggest that she is a classic UFO abductee. Additionally, Amy had been involved in meditation and other spiritual studies for years, which suggests that perhaps she had simply reached a new stage in her path, perhaps one that she did not expect. Yet her awakening did begin with a UFO-like dream experience. Perhaps the UFO dream was part of

---

[16] In subsequent conversations in preparation for this book, Amy indicated that her life has largely returned to normal. She did not in fact, pursue her plans to start a psychic study group. However, the same paranormal and metaphysical events have continued to manifest in her life in the same way, though to a lesser degree. Most importantly, Amy appears to have integrated these influences into her life.

the overall experience or perhaps it, too, was an effect of her awakening.

Perceivers and Experiencers

As others and I have continued to conduct research into the close encounter phenomenon, we have noted that this awakening effect frequently occurs at critical times in the experiencer's life. In addition, it often coincides with their emerging memories of UFO experiences. What does this timing mean? This still is not clear to me. However, as I continue to interview people who have experienced this sudden emergence I have noted a surprise...

Before I had begun my study and practice of hypnotherapy, I had interviewed many experiencers of paranormal events. Most of them described paranormal anomalies, psychic experiences and other strange events in their lives. Many had had multiple UFO sightings. As a result of their accounts and based upon the writings of other UFO researchers,[xxix] I had hypothesized that these witnesses were probably UFO abductees who simply did not remember their encounters. However as we later explored the experiences of some of them (often in deep hypnotic trance) I noted that at times we were completely unable to find instances of UFO abduction.[xxx] While they had indeed experienced awakening effects, the initiator did not seem to be UFOlogical. There are many ways to awaken.

Many studies indicate that unexpected psychic/spiritual emergence is a very common result of profound life-altering events.[xxxi] Events such as illness or trauma, near-death experiences, deep meditation, all seem to be possible initiators of an awakening. While the descriptions of each of these experiences can be very different, the effects are often similar. Many of these people subsequently seem to develop the ability to see UFOs and anomalies in situations where others around them often do not. Thus, I often refer to persons in which this UFO-seeing ability has been enabled as "perceivers."

Rosalie - a Life Time of Wonders

An example of a perceiver, who has had many UFO and anomaly sightings, apparently as part of a lifetime of paranormal sensitivity, is a woman I will refer to as "Rosalie." Rosalie is a pleasant, caring and very intelligent senior citizen who lives in an apartment building in the rural Midwest.

I had the privilege of interviewing her and her family and learned that Rosalie had many strange memories that stretched back as far as the 1920's. These included paranormal events, anomalies and UFO sightings. I first suspected that she might be an abductee. However as I conducted several hypnotic regressions, none of the classic abduction accounts emerged. Yet, while under hypnosis, she described in detail many encounters with the extraordinary.

Rosalie described how, as a child, she played with "little people" in the woods near her house. She described seeing helicopters during the early 1940s, although machines matching her description apparently did not exist until many years later. She described balls of light and a host of other mysterious psychic perceptions. Yet we did not find any UFO abductions, per se. Thus, Rosalie is someone whom I would regard as a perceiver. While she has had a few UFO sightings, they seemed secondary to her overall awakening.

"Anne" – A Spectacular Sighting

Another perceiver whom I was privileged to interview extensively was "Anne." Anne is a woman in her 40's who lives in Minneapolis. She has a long history of interest in eastern spirituality and metaphysics, meditates regularly and has had numerous mystical experiences. Over many years, Anne has steadily developed a high degree of psychic ability. Like Rosalie, few of her experiences have directly involved the UFO phenomenon. While she has become steadily more "awakened," this has been a gradual growth process.

While Anne has had few actual UFO events in her life, she was witness to one very spectacular sighting, the South

Minneapolis blackout sighting (described in chapter 1). In this event, Anne saw, to her surprise, a large UFO hovering only a few hundred feet above the building tops only a few blocks away. The object was positioned in such a way that everyone should have seen it. Yet Anne seemed to be the only one aware of it.

Many such cases exist in my files and those of other researchers in which a few apparently sensitive witnesses are able to discern the presence of a UFO while others might not see it. In many of these cases, the witness had previously experienced the awakening phenomenon to a high degree. However, not all of these perceivers are abductees. Rather they experience the UFO phenomenon primarily because their heightened sensitivity allows them to perceive it.

Contactees, Already Awakened UFO Experiencers?
In yet another group, we find the contactees.[xxxii] Far more involved in the UFO phenomenon than perceivers, contactees generally describe close encounters in which they are voluntary participants. In most cases, these events seem to be positive in nature, involving extraordinary but generally benevolent entities. These experiences tend to enhance the psychic/spiritual path of the contactee. Also, along with the overt contact experiences, contactees frequently describe communication that is apparently channeled in nature, suggesting increased psychic/paranormal perception.[xxxiii]

Since their experiences are largely positive in nature, few contactees have sought hypnotherapy. Thus, I have generally not been a party to exploration of their experiences at the deep subconscious level. Yet as a MUFON investigator, I have interviewed many contactees who live in the Minneapolis/St. Paul area. Most have indicated that contact was simply part of a long path of metaphysical growth and exploration. Many of them tell me that their life path had led them to be more sensitive to the extraordinary and thus eventually to contact. Thus, in this case their UFO-related events seem to be an effect rather than the cause of their awakening.

A Synergy of Awakening and Encounters

From the cases in my files and the files of other researchers, as well as in much of the literature, it appears that the UFO phenomenon is one of many possible initiators of unexpected psychic/spiritual awakening. Many find that the abrupt, often-frightening experiences of UFO abduction force them to begin a path of awakening. However we also find that many other important life episodes, often unrelated to the UFO phenomenon, can have the potential to initiate a person down this same road. Furthermore, I noted earlier how perceivers of the extraordinary often tend to be more prone to UFO sightings. In many cases, they are not UFO abductees but simply sensitive bystanders. They seem to be able to see and experience anomalous objects that the "unawakened" may not observe.

For still others close encounter experiences may be a result of, rather than the initiator of, a path of psychic/metaphysical awakening. Generally contactees, they report mostly positive experiences with benevolent entities.

Some authors describe how there appears to be an increasing spiritual and psychic awakening occurring among aggregate of humanity. In her book, *Conscious Evolution*,[xxxiv] Barbara Marx Hubbard cites studies showing this phenomenon and suggesting that this is approaching a threshold point.

Other researchers describe how there appears to be an increase of close encounters.[xxxv] Thus, we can ask whether a possible upsurge in reported encounter events might be associated with an increased number people with growing psychic abilities.

What is the relationship of the UFO phenomenon to that of awakening? After our examination, we find this relationship to be as ambiguous as most other aspects of the UFO phenomenon. Sometimes awakening is the cause, sometimes the effect. In still other cases, the two seem to occur in parallel, in almost a lockstep manner. They appear to be related yet the relationship is not clearly defined. And so, like most other aspects of UFO and anomaly research, the phenomenon of unexpected awakening is a mystery that beckons us down that long winding path, which in the end leads us to still more mystery.

Calling on Extraterrestrials - The Proactive Views of Lisette Larkins

Some experiencers, as they become more aware of the phenomenon, seek to take a more active role in their encounters. I have noted that they often seek out other experiencers to form communities, attend meetings of groups such as the local MUFON chapter, or become involved in similar activities. In addition, some overtly attempt to communicate with the visitors themselves. One author, Lisette Larkins, offers us some interesting views on the possibilities of such proactive contact.

In her book, *Calling on Extraterrestrials*,[xxxvi] Lisette Larkins offers a new view of the close encounter experience. Many other books study the phenomenon from the outside, describing the experiencer as a victim. They try either to analyze or to debunk the phenomenon. However, in her book, Larkins describes the CE4 experience in a unique way. This book views encounters from the perspective of proactive contact, working with the phenomenon on her own terms. To Larkins, being an experiencer offers the possibility of taking the initiative in a relationship with the visitors and exploring what this could mean in one's life.

Larkins begins the book with an introductory section describing how her adventures with contact began - a narrative that sounds very much like most other abduction accounts. She describes how she, like many other abductees, was isolated from others (including her husband) because of her interactions with the phenomenon. She then describes how she began to take hold of her relationship with the visitors and to turn it into a more meaningful, positive interaction.

In *Calling on Extraterrestrials*, Larkins describes an approach that she says can open the door to personal contact. She recommends a way to choose contact and to assert the intent for the form in which it should occur. She suggests that in order for this to work it is necessary to tame the fear and to release and forgive issues, both with the phenomenon and with other humans, which would prevent constructive contact.

Larkins next describes ways to expand your worldview to accept contact. She discusses the need for discernment to understand and be proactive regarding the contact that you wish to receive.

In the final sections, Larkins further elaborates on how to be proactive and provides steps for personal transformation. She suggests ways in which the relationship between the experiencer and the phenomenon can evolve and how the experiencer can grow with the phenomenon. In the end, she says that the phenomenon is one means of transformation by which the experiencer can grow beyond the bounds of his/her former self.

Larkins offers a largely positive view of the phenomenon and many abductees describe much darker experiences than what this book portrays. However, Larkins states that "you call the experience and then experience what you called," - meaning that you experience contact according to your own beliefs and expectations, even if such contact is voluntary, as in her case.

Larkins further focuses on personal spiritual growth as a way to foster the contact experience and much about her approach resembles other meditative/spiritual paths available to the seeker. Yet this reinforces the point that the visitor experience is just one of many possible avenues for personal growth. Like other spiritual paths, Larkins' proactive approach to the visitor phenomenon beckons us to awaken.

Other Human-Initiated Contact

Many other publications and groups also explore the idea of human initiated contact. One of the best case studies of humans actively participating in interactions with the visitors is the book *CE5*, by Richard Haines.[xxxvii] This book presents a large number of cases in which humans have somehow initiated contact, with a resulting response by the UFO phenomenon. This may be as simple as witnesses blinking their car lights and observing a UFO blink in response, or it may be as complex as multiple participant visitor encounters.

In another deliberate effort at contact, the organization CSETI,[xxxviii] founded by Dr. Steven Greer, maintains an ongoing program entitled "The CE5 Initiative." This effort is specifically intended to initiate contact with the visitors. At the core of the CSETI strategy is a protocol for contact using a form of meditation, which Dr. Greer entitles "Coherent Thought Sequencing." In regular outings during the early to mid 1990s, CSETI contact teams used this methodology to attempt to "vector in" ET craft and initiate contact. The CSETI literature claims that they have had some significant positive results. The interested reader is encouraged to consult the CSETI website and/or read Dr. Greer's book, *Extraterrestrial Contact: the Evidence and Implications*,[xxxix] which provides the fascinating details of this effort.

Jenny - It's about choice: a lesson from the visitors

I have observed that, as well as experiencing the more benevolent side of contact, experiencers are often confronted by the darker side of the phenomenon. While conducting recent interviews and regressions, I have observed the dark side at work in all of its sinister intensity. In many such cases, the character of the phenomenon will change for a particular period in the experiencer's life. The "normal" lifelong relationship, which the experiencer has developed with the visitors, particularly with key entities, suddenly takes on a darker tone. Sometimes the same entities are involved but, in other cases, different entities appear to be at work. A more sinister side of the phenomenon, more frightening, more exploitive, or simply more evil, may replace the relatively familiar events of earlier in life.

One woman whom I will call "Jenny" has had a lifelong history of close encounters. Throughout much of her early life, she had experienced encounters with a being that I will refer to as "the Captain," a gray who seemed to be her handler and keeper. This being was relatively neutral in character but seemed to have at least a functional interest in Jenny's well-being. He treated her with respect and as a result, she did not fear him.

Then, at a point in her twenties and early thirties, her relatively positive encounters with the Captain were suddenly replaced by encounters of an entirely different character. She found herself, instead, abducted by beings who handled her much more roughly than the Captain and his "away team." These other beings had little care for her well-being. Their entire focus was on conducting painful medical experiments on Jenny's body.[17]

These negative encounters were painful, frightening and intrusive. They resulted in a steadily increasing level of fear (post-encounter stress) being manifest in Jenny's life. On at least six

---

[17] Jenny noted that these experiments did not seem to make sense. They seemed more designed to inflict pain than to gain knowledge or accomplish a reproductive or medical purpose.

88

distinct occasions over several years, Jenny found herself cowering in the back corner of her walk-in closet, crying. Each time she eventually composed herself, crawling back into bed. But with each event, the fear deepened.

Then, as quickly as they began, the negative events appeared to discontinue. Yet even ten years later, the traces of fear remained deeply buried within Jenny's psyche. This fear suddenly manifested in all of its brutal force during one hypnotherapy session. Halting the regression, we shifted our focus to fear management. After some brief but very effective hypnotherapy sessions, we were able to resolve the fear by addressing some deeply ingrained issues. Once these were addressed, her fear reactions were greatly reduced. We were then able to proceed with the regression.

In spite of the fear management therapy we had conducted, the events of the abduction, which Jenny was reliving, were still unpleasant. However, we were able to examine the details of the medical procedure, which the aliens had conducted upon her. By understanding the details of the procedure, we were better able to understand the motivations behind the phenomenon and the meaning of the encounter to Jenny. This occurred in a flash of insight during the depths of the encounter, as she suddenly realized what the visitors were doing and why...

During a regression, as she continued to describe what appeared to be a very unpleasant procedure, Jenny expressed anger and disdain for the beings who were conducting the experiment. At that moment, I asked her to pause (freezing the event on a given frame, much like a paused video or movie frame) and to imagine that the leader of the aliens was sitting in a nearby (but still comfortably distant) chair. I invited her to say whatever she wanted to the being (with some clients, I have found that this can result in some colorful language).

Sometimes referred to as "chair therapy," this therapy can be very effective in fostering forgiveness and healing of issues remaining from earlier painful relationships. In this case, Jenny let the leader of the evil aliens verbally "have it" in no uncertain terms. She then told the leader that, just as she had before, she was denying him permission to continue his experiments. She was saying (and actually had said at that time) "NO!"

I then asked Jenny to imagine what the leader was saying in response. The result was that the leader congratulated Jenny. She had finally "gotten the lesson." The lesson was about choice and Jenny had made that choice. She realized that she is a free soul with self-determination. She could choose to deny the beings control over her.

As a result, she was able to stop these darker aliens from conducting their unpleasant procedures. Thus, this particular dark-side event seems to have been the last one. From this point on her relationship with the phenomenon would again be on a more level playing field.

The lesson, as Jenny described it, was about free will, or as Jenny said to me afterwards, "It's about choice." We are being given the opportunity to learn about the light and the dark side of the phenomenon and to choose the side with which we wish to associate. Since our relationships are often reflections of our own self-image, in so doing we learn to enhance that side of ourselves.

Contrary to popular conceptions that the phenomenon is evil, the lesson taught to Jenny is that we are free will beings living in a free will Universe. Like most challenges, this series of frightening, negative encounters have had a beneficial effect - to spur on Jenny's own growth. Over the ensuing years, Jenny has become a deeply spiritual being, a gifted and empathic healer with a caring heart. Yet she has also developed an inner strength that only comes through the forge of torment and challenge. Thus, in its own way, the visitor experience, complete with its dark side of fear and pain, is perhaps just a footstep on Jenny's path over the Cosmic Bridge.

Personal Spectrum of Contact

In the literature, we see a number of paradigms or models of contact. Some claim that the visitors are evil and are exploiting us for reasons only they understand. Others assert that they are good and are here to enlighten us, or that they are somehow divine in nature and are here on some type of mission from God. These are just a few very broad descriptions of the many possible models of our interactions with the visitors. Are all of these true? Are any of these views mutually exclusive? How can we place these assorted models into perspective?

In the book *Healing Shattered Reality*, by Bryant and Seebach, the authors describe a spectrum of views based upon the personal paradigm of the person being contacted. They present a range of scenarios, from (potential) scientific recognition and study, through abduction and initiation and finally spiritual enlightenment. The book describes these as simply four points on a continuous spectrum of contact scenarios. The authors further state that the contact occurring to each experiencer is aligned with their personal paradigm along this spectrum.

1. Recognition:

This first phase of the paradigm spectrum is the rational, scientific stage. At this stage of the spectrum, a person can accept the potential existence of extraterrestrial life and intelligence, as long as ET remains "out there." While one is in this paradigm, contact can occur only via means of scientific speculation - or via SETI or other distance communication models in accordance with presently understood physical laws.

This appears to be the present state of thought within the western scientific community. Scientific study of biology and biochemistry are acceptable, along with potential search for distant ET civilizations. However, the possibility of ET visitation to Earth remains unacceptable. Thus, in the present day there is little likelihood of official acceptance of possible ET visitation.

This degree of non-acceptance creates the high level of discord, which currently exists between the SETI and UFO research communities. While most members of the SETI community are at this "Recognition" stage of the spectrum, most members of the UFO community are at different stages. Neither community accepts the fundamental paradigms of the other, thus preventing common viewpoints and perhaps even meaningful dialog.

An extension of the "Recognition" stage as described above is the "Nuts-and-Bolts" UFO model. In this paradigm, we allow that an ET civilization might have full interstellar travel capability. Thus, we arrive at a paradigm of physical ET Visitation in which it is recognized as fully reasonable that "they" might be visiting Earth. For many in our society this appears to be reasonable as long as entities either are not presently contacting humans or are contacting "someone other than me..."

2. Biological Intervention:
The second phase of the paradigm spectrum is what Bryant and Seebach refer to as the "Biological Intervention" scenario. This is the UFO abduction model in all its glory, what I refer to in chapter 2 as the "standard model" of UFO encounters.

In this model, the purpose of contact appears to be genetic in nature. It has a strong emphasis on the taking of genetic material from human abductees and using it to breed an intermediate hybrid race (for purposes unknown to us). The relationship between the experiencer and the visitors seems to be one of "lab animal" to "scientist." There is a cold, clinical essence to it, in which the visitors seem to have little interest in anything other than conducting medical experiments upon the abductee.

In this phase of the paradigm spectrum, the motivations of the visitors may be benign, benevolent or malevolent. It largely depends upon the point of view of the experiencer and/or UFO researcher. Yet, as we have discussed earlier, the UFO encounter seems to be a mixed bag of positive and negative experiences.

3. Initiation:

In the third phase of the paradigm spectrum, many experiencers have referred to ET contact as being a form of initiation. I have noticed that, in the descriptions by some experiencers, there does appear to be an initiatory aspect to the phenomenon. To many experiencers one effect of their interaction with the phenomenon appears to be the introduction of - and initiation to - some form of "higher perspective." Some experiencers state that it is as if they are being inducted into a "special order" with an "important purpose."

We earlier noted how a large number of experiencers describe having such a special purpose or mission. It is not clear what the purpose of such a role is, either for the experiencer, or for the visitors. Thus, this aspect contains a tremendous degree of mystery.

4. Self Awareness:

Alice Bryant describes the top end of the paradigm spectrum [my paraphrase], as that of spiritual awareness and enlightenment. In this model, the phenomenon is present, at least in part, to bring the experiencer - and all of humanity - to a higher level of spiritual understanding. This takes the experiencer to a new and/or expanded level of freedom, awareness and insight. In this view, the phenomenon takes on a more positive tone and conveys the sense that somehow "it was worth the cost" of whatever earlier difficulties the experiencer may have undergone. Most contactees appear to have this view of the phenomenon.

Bryant and Seebach describe these as four primary phases of contact but indicate that there may be many more and/or many combinations of these. In a manner similar to our earlier discussion of experiencer perspective (chapter 4 and earlier in this chapter), Bryant and Seebach state that ultimately, the nature of the relationship between the experiencer and the phenomenon seems to be dependent upon the perspective of the experiencer. In the words of Lisette Larkins, "you call the experience to yourself and then experience what you called."

Other Extraordinary Experiences

In preparation of my article "The Indigo Hypothesis"[xl] for the MUFON Journal, I read a number of new books on the extraordinary experiences of children. As I worked through the material on this topic, I noted that much of the literature focused on children's Near Death Experiences (NDE). I thought about this in the context of the UFO/close encounter phenomenon and began to wonder about the relationship between the NDE and the UFO encounter? How do they compare in their abilities to affect the spiritual emergence of the experiencer?

Much of this literature, especially writings by PMA Atwater[xli], focuses on the near death experience (NDE) and how it affects one's life. It describes the adjustments of both childhood and adult near-death experiencers to their experience. It also describes how children who undergo near-death experiences appear to have more difficulty than do adults in adjusting to their return to "normal" existence in the earth-plane. They seem to take many years to adjust, often remaining separate from childhood social activities, non-participative in school activities, etc. They often tend to have deep spiritual wisdom, psychic abilities, empathy and higher than average intelligence. Many describe how they live partially "on the other side" and that in their view, there is a gap between them and the physical world.

As we described earlier, any event of the fourth kind comprises a reality transformation, where the experiencer becomes fully involved in the reality of the phenomenon. Two different types of events of the fourth kind, which can deeply affect the experiencer, are the Anomaly (AN4) and the close encounter (CE4).[xlii] Anomaly events include those such as out-of-body experiences and near-death experiences, whereas CE4 events are primarily those of alien abduction.

We can note both similarities and differences between such events. At first glance, the two phenomena appear to be very different. The classic alien abduction event is generally intrusive and disempowering. It appears, in general, to be a non-positive event. The relationship between alien and human is frequently described as cold, uncaring and clinical, whereas the near death experience is one in which the experiencer finds him/herself in a realm of loving beings, in the presence of God, angels, etc.

However, on occasion, there can be a crossover between these two phenomena. In the Kenneth Ring's book, *The Omega Project*, he notes that descriptions of Near-death events occasionally involve UFO, alien or extraterrestrial themes. Similarly, I have noted that some UFO encounters involve past lives; God, Jesus and/or other religious figures; spirits or metaphysical entities; or related themes.

Just like near-death experiencers, some UFO experiencers describe some of their encounters as more spiritual experiences, with beings imparting wisdom, psychic abilities and a special purpose. In his book, *The Andreasson Affair*,[xliii] Raymond Fowler describes an abduction experience of Betty Andreasson-Luca. During this experience, she is taken into an alien realm, where she encounters a powerful spiritual being. Betty concludes that this being was actually a manifestation of God.

In many cases, we see that both experiences can have similar after effects. As we described earlier, many experiencers, of both UFO abduction and near-death events, describe how they subsequently feel special, that their home is "out there" and that they sometimes feel out of place in human society. Similarly, there is the sense of acquiring new knowledge and perspectives, which others cannot understand. There is often a tremendous degree of adjustment required when they first become aware of the dynamics of the phenomenon in their life.[xliv]

PMA Atwater describes how those who have near-death experiences often require a lifelong coming-to-terms with their experience. Just as UFO experiencers occasionally describe the sense that their home is out among the stars, there is often the desire by NDE experiencers to "return home" after having briefly experienced the loving beauty of the after/inter-life.

Earlier in this chapter, we noted that one phenomenon common among many types of experiencers is that of spiritual awakening. Both UFO abduction and near-death experiences seem to engender the psychic enhancements, spiritual shifts and changes in the personal paradigm of the experiencer. As Barbara Harris Whitfield describes in *Spiritual Awakenings*, such events can occur because of any number of life changing events: Close UFO encounters, near-death experiences, health or life crises that do not actually involve extraordinary reality transformations, even emotional stress or deep meditation. All appear to be potential catalysts for personal psychic and spiritual transformation. Subsequently, the experiencer seems to have enhanced psychic ability and with it, a subsequently enhanced tendency toward further extraordinary events.

In this sense, though the detailed description of the experience may be different, near-death experiences and UFO abduction events appear to be related experiences. Thus, we can ask if there may actually be a continuum of experiences, ranging from the purely physical nuts-and-bolts UFO encounter, to the purely spiritual/metaphysical experience.

A Bridge between Worlds

As we noted earlier, one effect of the phenomenon on the lives of experiencers is the apparent perception that they do not belong here. Their home is "out there" or "on the other side." They sometimes describe having difficulty functioning in our mundane earth-plane society. It has been described to me how this can sometime lead to depression, non-participation and withdrawal. Carolyn, Evelyn, Amy and most of the other abductees we met earlier have described this walk between worlds.

David Jacobs, in his book, *Secret Life*,[xlv] elegantly portrays this theme. He describes how the abductee seems to lead a double life, both in the world of the mundane and the world of the phenomenal. For most, life in the phenomenal realm is a closely guarded secret. It forms a wall of isolation, separating the experiencer from the everyday world and makes the close encounter experience one of the biggest hurdles in life that I can imagine.

In Raymond Fowler's book, *The Synchrofile,*[xlvi] he describes his own experiences of metaphysical, synchronistic and close encounter events, all intimately intertwined. He describes how he has had both UFO and psychic experiences throughout his life. He then speculates on their relationship and its implications and compares his events to those of other experiencers. He, too, concludes that there are two separate worlds in the life of an experiencer: the everyday world and the world of the phenomenon.

Many experiencers with whom I have worked have described this same sense of having two separate lives. The phenomenal world seems to be a world in which they are heavily involved, yet of which they have little overt awareness. Yet this hidden world secretly seems to dominate the experiencer's life.

We can observe the effects of close encounters on the experiencer's life - yet, as Budd Hopkins has stated, "We live 99.9% of our life in this world."[xlvii] Thus, we must ask how we can forge a bridge between these two realms.

Ultimately, for the experiencer, a goal in the healing of shattered reality[xlviii] may well be to function in this world while being aware and integrated with the 0.1% of life spent within the phenomenal realm. For each experiencer this is the challenge - the forging of his/her own Cosmic Bridge.

The Awakening of a New Perspective

Are the visitors good or evil? I have come to ask, "Good or evil for whom?" The answer that I find is that our interactions with them seem to transcend that simple question.

We have seen in this and previous chapters how the close encounter phenomenon can represent forces of darkness or of light in the life of the experiencer. Based upon one's perspective and perceptual filters, what the phenomenon brings can be gifts or curses, fear or enlightenment. We have also seen that perhaps, through healing modalities such as hypnotherapy, an experiencer can come to terms with the phenomenon on neutral ground, less as a victim and more as a participant. In the light of understanding perhaps one can see close encounters as the mixed bag they are, some unpleasant and some enlightening. In the end, rather than either good or evil, perhaps the phenomenon simply is what is.

We have also seen how interaction with the phenomenon indelibly changes the experiencer in many respects. Depending upon our perspective, we can see this as a form of trauma, or as a form of learning which fosters a shift in metaphysical and spiritual perspective. We have noted this shift along with the accelerated development of psychic abilities - an awakening.

We have further discerned that psychic and spiritual awakening seem to have a synergistic relationship with the close encounter phenomenon. While UFO encounters can induce a shift in metaphysical perspective, awakening by any influence seems to make one more sensitive to the UFO phenomenon. There appears to be a positive feedback effect, which enhances the power of the phenomenon in the life of the experiencer.

In this chapter, we have looked at the themes and trends of the contact relationship that permeate the life of the experiencer. We have noted how the phenomenon grows with the experiencer and transforms his/her life. We have continued to expand our paradigm of the deepening relationship between humanity and the phenomenon, taking us one more step down the path of contact.

In the end, we begin to find that the phenomenon awakens each person into an individual relationship. The nurturing of such a relationship seems to bridge the everyday life of the experiencer with the realm of the phenomenon. For better or for worse, it forces each into the building of their own Cosmic Bridge.

Chapter 6 - Close Encounters and Human Genetics

The encounters continue but the mystery remains. In chapter 2, we developed the details of a typical alien abduction, including the capture and medical examination by entities from elsewhere. In the chapter 4, we saw how, to the experiencer, the abduction experience is a deeply personal one. Yet the biggest mystery remains - what are the visitors actually doing? What is their purpose for being here?

As more abduction accounts began to surface, researchers noticed many common themes in the procedures abductees described. It soon became apparent that human reproduction was a key focus of the visitor interaction. In his books, *Missing Time* and *Intruders*,[xlix] Budd Hopkins was the first to note this as a core of the phenomenon. Hopkins and Jacobs have since established that the visitors seem to have tremendous interest in human genetics.[1]

What is the purpose of this focus on human procreation? In this chapter, we will examine this genetic aspect to the close encounter scenario. In subsequent chapters, we will see how it may be one of the key elements of the Cosmic Bridge.

Hybrids and missing fetuses

Whatever the visitor agenda may be, descriptions by abductees strongly suggest that the visitors are conducting some form of reproductive procedures. Nearly every experiencer has described the harvesting of genetic material, while female abductees have often described the implanting and subsequent extraction of apparently-hybrid babies. Many have experienced subsequent child presentations,[li] in which they are shown babies and told that the children were theirs.

Similar to accounts in *Intruders*, several experiencers with whom I have worked have related heart-wrenching accounts in which they have been impregnated or otherwise implanted with embryos. They carry these embryos partially to term, only to have their pregnancy "disappear" a few months later, well prior to the expected time of delivery.

A typical account[18] is the story of "Sandra." Sandra had undergone a series of abduction experiences during her late teen and early adult years. Most of these appeared to involve medical examinations with a focus on her reproductive system. Several years later, while lying deep in trance in my hypnotherapy studio, she recounted some of these events.

Sandra described how, subsequent to one abduction experience, she learned that she was pregnant. At a relatively early point in her pregnancy, while lying in bed, she suddenly found herself again surrounded by small gray beings. A typical abduction scenario ensued in which the beings took her from her room into what she described as a UFO. They took her to an examining room, where strange looking "doctors" with solid black eyes entered. She described how they "removed something," which she suddenly, with a heart-wrenching cry, described as "My baby!"

---

[18] This account is actually an aggregate of several similar events. In each case, the details are obscured to protect the anonymity of those involved.

Sandra noted that, on closer examination, these "doctors" were short and had pale skin. They had no hair and had large black eyes. They appeared to be the usual gray beings associated with "standard" UFO abductions.

Many abduction accounts have a theme typical of Sandra's story. A number of elements within such descriptions convey this same genetic/reproductive theme:

1. Extraction of sperm and ova from the bodies of experiencers: This appears to be a relatively standard procedure, universally described as part of the abduction experience.

2. The apparent crossbreeding or genetic manipulation of the extracted material to create some form of human-alien hybrid embryo,

3. Embryo implantation and the subsequent removal a few weeks to months later, of a partially developed being

4. Incubation of the extracted fetus: Many abductees describe being shown a room containing incubation tubes, which apparently contain hybrid fetuses in varying states of incubation.

5. Child presentation: Some time subsequent to the fetal extraction, the experiencer is shown a child whom they are told is their own. This is described by a high percentage of female abductees.

A case that further illustrates these events is that of the experiencer introduced earlier as "Carolyn." She has had numerous UFO encounters and feels that she has at least one "star child" who is "out there somewhere."

Carolyn and her husband live in a rural wooded setting, near a moderate sized town in the upper Midwest. At one point, they had an up-close UFO sighting, which they subsequently realized was probably a UFO abduction. Shortly after their encounter, Carolyn learned she was pregnant and the happy couple began looking forward to becoming a family.

At a point several months after their encounter, Carolyn was again abducted. Subsequent to this, she discovered to her horror that she was no longer pregnant. Although it was relatively late in her pregnancy and was not aware of any actual miscarriage event, no baby was present.

In the next few weeks, Carolyn began to have nightmares of UFOs, aliens, hospitals and bright lights - dreams frequently associated with UFO abduction. In the next several months, she also began to remember more details of her initial UFO abduction, which had apparently occurred at the time of their sighting.

A couple of years later, Carolyn was again abducted. This time, she was shown a child who had the appearance of a human boy approximately two-years in age. Carolyn "knew immediately" that this was her child. From that point on, Carolyn has considered him to be her "star child."

Incubation Tubes

A large number of experiencers have described rooms filled with incubation tubes containing fetal-appearing beings in various states of incubation.[lii] Representative of such cases is the description by an experiencer whom I will refer to as "Susanne."

Susanne had recently undergone a missing-pregnancy experience similar to that of Carolyn and Sandra, in which beings had apparently removed a fetus from her. In a subsequent abduction a few weeks later, she was taken to the location in which these beings apparently complete their gestation. She was ushered into a room containing many large glass tubes. In each of these tubes, she observed a fetus-like being similar to the one removed from her. She was told about these beings, including that one of them was her child. Susanne does not feel that she has ever seen this being again but is sure that she also has a "star child" out there somewhere.

Genetic Re-Engineering

As we have seen in this chapter, study and manipulation of the human genome seems to be important to the visitors. Could it in fact be at the core of the abduction phenomenon? In their books, Budd Hopkins and David Jacobs strongly suggest that genetic manipulation is the primary reason for abduction. Hopkins builds an excellent case for the existence of this agenda in his book, *Intruders*. Jacobs further builds the case that there is a coherent purpose to this process.[liii] Many others have elaborated on this abduction/hybridization scenario[liv] with various interpretations and conclusions.

In *Sight Unseen*, Budd Hopkins and Carol Rainey build a fascinating case for the plausibility of such a quasi-hybridization endeavor, describing in considerable detail how it can occur. Such a process does not seem to involve direct crossbreeding of species, but rather transgenics - in which selected genes encoding characteristics in one species are transplanted into the genome of another.

Hopkins and Rainey cite experiments in which scientists have already accomplished the transplantation of genes from one species to another.[lv] [CL Note: On the internet,[lvi] I found several descriptions of research efforts involving the transplantation of genes from one species to another.] Hopkins and Rainey theorize that such transplantation is being done to the human genome, for purposes known only to the visitors.

This leads us to the speculation that DNA, either synthesized or somehow altered, could be combined with that of humans to build the genetic blueprints of the hybrids commonly described in abduction accounts.

Alien Encounters in our Everyday World

What can we discern about the end product of such genetic manipulation? Who or what beings would result from such actions? Would they be limited to the alien realm or might they, on at least some occasions, exist in our everyday world? Might such human aliens even move among us?

Some of the strangest aspects of the UFO and close encounter phenomenon are reports of encounters with entities that appear nearly human but have some distinguishing features placing them in the alien realm. While such beings are often seen aboard a UFO during close encounters, experiencers occasionally also report encounters in our every-day reality.

The Strange Beings in *Sight Unseen*

Budd Hopkins carries the theme of genetic manipulation further, describing a number of encounters between experiencers and the apparent result of this hybridization process - humans who seem to have alien characteristics.

A fascinating collection of such encounters can be found in his book *Sight Unseen*,[lvii] written with Carol Rainey. This book describes encounters with beings that are apparently human, but which appear to have extra-normal characteristics. They are apparently associated with the UFO abduction phenomenon, yet reside within our everyday world.

One case in *Sight Unseen*, which merits special attention, is that of an abductee, who Budd refers to as "Ann-Marie," and her strange companion "Mr. Paige." Budd describes Ann-Marie as a warm, caring woman and a life-long abductee. He describes how she, as well has her brother "Peter," have had unexplained experiences throughout their lives, especially during childhood. Mixed into this scenario is the account of their strange acquaintance, Mr. Paige, who resided with Ann-Marie's family for nearly a decade.

Mr. Paige is described as old and wise. His features were odd, with a large head and long chin. As Anne-Marie described him, Mr. Paige seemed quite extraordinary. This strange man seemed to be without friends, family, or possessions. Budd describes him as having no apparent credentials such as driver's license, credit cards, etc., leaving the impression that he barely registered on the radar of society. The only indication that he ever gave of his origin was an ambiguous "up north."

The apparent alien nature of Mr. Paige was enhanced as Ann-Marie remembered strange writing, which he would pen as he wrote letters (and even a book). She described the writing as nearly incomprehensible, with strange symbols and script. Yet at the same time, when Ann-Marie read letters he wrote to her in later years of their acquaintance, she seemed fully able to understand them. Budd describes how this writing was similar to alien writing seen by other abductees during close encounters.

Mr. Paige was a long-term friend of the family and became quite close to the children. In subsequent memories however, Ann-Marie began to recall scenarios in which she and Mr. Paige would take "nature walks" in the forest. She later came to see these as settings for UFO abductions. On multiple occasions, she described such half-remembered experiences in which something strange happened.

For many years, Ann-Marie had no conscious memory of abduction events. However, as time went on and she had begun to meet and work with Budd, she started to remember her encounters. It was apparent that Mr. Paige was somehow associated with the UFO abduction phenomenon.

Who or what was Mr. Paige? Was he a transgenic, partially alien being? What was his real relationship to the normal world of humanity? Furthermore, what was his relationship to the world of the UFO phenomenon? As Budd describes in *Sight Unseen*, the true nature of the strange being that is Mr. Paige is a mystery.

Similar Local Cases

Several cases exist within the case files, both mine and of other researchers, describing beings similar to Mr. Paige. In one case,[19] an experiencer whom I will refer to as "Paul" described a strange "friend" whom at the time, he viewed as a mentor or leader. Paul's friend had a magnetic charisma and exerted what Paul described as a "psychic" hold on others. Many psychic and paranormal events occurred, nearly all somehow associated with Paul's friend.

Paul also described how, on more than one occasion, his "friend" escorted him to meet (and be abducted by) the aliens. In one key event, Paul's friend showed up unexpectedly and told Paul to follow, which he did. The friend took Paul into a building where an alien encounter occurred.

On subsequent investigation, I learned that many details of the account were real. However, the building that Paul described was probably the screen memory of a UFO, the location of the abduction.

In another case, an abductee described how she had seen a strange man on multiple occasions, each immediately prior to an abduction event. Later, while in hypnotic trance, the abductee stated that this stranger was apparently involved in her encounters, though she could remember little of her interaction with him.

Was this man somehow involved in bringing her to the encounter scenario? Was he, like Paul's friend or Anne-Marie's Mr. Paige, an apparent human-alien, perhaps working in league with the phenomenon?

---

[19] The case of "Paul" is actually an aggregate of several similar cases.

Questions and Hypotheses

If we take these witness accounts at face value then who or what are these humans that also seem "alien" in nature? Are they physically real? Or might there be other ways of understanding them? One hypothesis advanced by David Jacobs in his book *The Threat*, is that they are indeed physically real. Jacobs suggests that they are the late-stage products of the breeding/hybridization project associated with UFO abduction. He further suggests that they are somehow assimilating into human society, for purposes that he believes are less than benevolent.

Jacobs notes that these beings seem to reside in the realm of the phenomenon, but that they sometimes go on "away missions" into the human realm, what he refers to as "Independent Hybrid Actions." Yet, some of the cases we have studied in Minnesota MUFON, and that of Budd Hopkins' "Mr. Paige," suggest that the visitors might maintain a longer-term presence within human society.

Ultimately, we can ask if these hybrid beings are the result of the transgenic process Hopkins and Rainey describe in their book, *Sight Unseen*. Are such beings just one step in the integration of alien genetics into human society? Are they truly humans, yet also bearing alien genome - or are they something else? - Truly, human-aliens...

A Genetic Agenda?

While we can speculate on possible genetic manipulation, we must accept that we understand little about its possible nature or purpose. David Jacobs proposes the most concise hypothesis, building the sobering case that there is a dark but unknown purpose behind it. However, Budd Hopkins states that their purpose is unknown and at the end of *Sight Unseen* expresses "cautious optimism."

When taken on its own, the "hybridization" process that most researchers describe does not clearly suggest a defined end-point or goal. Yet, when taken in the context of the other aspects of the phenomenon: interactions with "human aliens," the psychic/spiritual aspects of close encounters and the warning messages which seem to be imparted during abduction experiences, perhaps a more coherent picture begins to emerge.

One hypothesis is that the purpose of a hybridization process might simply be to breed a "better" human. The literature describes one possible representation of such an "advanced human" as the "Indigo child." We can speculatively ask whether the Indigo children are in some way a product of the hybridization process. I refer to this idea as the "Indigo Hypothesis" and ask whether this might be a key aim of the visitors. In chapter 9, we will discuss this and other hypotheses in more detail.

In the next chapter, we will begin to expand the individual relationship with the visitors, to examine their relationship with humanity. What is their objective and how do they intend to accomplish it? How do the abduction and hybridization process and the transgenic beings that apparently result help to accomplish this agenda? Is such an agenda for our benefit or otherwise? Are these beings one step in the integration of humanity with the cosmic community? If so, then perhaps such an agenda is a key step in the building of the Cosmic Bridge.

Chapter 7 – Classical Models of Contact

Ping, Ping, Ping... It arrives from far beyond the arrays of antennas and beyond the arsenal of scientific instruments. From somewhere among the stars comes a signal. Is it a message meant for us? Is it a deliberate sign from a radio beacon somewhere out there, or did some lucky human scientists simply hear a bit of radio chatter amongst the stars? In my imagination, I can picture a researcher, something like Jodie Foster in the movie *Contact*, happening across such a signal in the middle of the night.

Or can I? Is this scenario, so dear to the SETI community and now relatively well accepted in the world of mainstream science, an accurate depiction of future contact with the extraterrestrial community? What would contact on a planetary scale mean and how would it likely occur? Or, as we have suggested in previous chapters, is contact occurring right here on Earth in the present day?

In previous chapters, we looked at the experience of individual contact with visitors from elsewhere. We saw how we could view contact as a relationship between the experiencer and the phenomenon. From this point on, we will look at contact from the point of view of society as a whole. Why are the visitors here? What are they doing? And what does this mean to us?

In this chapter, we start to develop the big picture. We will first look at some of the classic paradigms of ET life and intelligence - and the more orthodox ideas of alien arrival. How does the traditional scientific community view potential contact? Up to this point, how has the UFO community, itself, envisioned meeting the visitors?

Finally, we discuss potential scenarios - and perils - of contact with the cosmic community. This sets the stage for our ultimate goal, to build a synthesis of ideas - a hypothesis of how contact may be unfolding in our time.

ET Life and Intelligence Paradigms

To discuss the topic of contact between humans and extraterrestrials we must first consider our partners in contact, the ETs. What would they be like? How would they think? How would they act?

We have already seen that there are many different views as to what constitutes a real or potential extraterrestrial presence in the universe.[lviii] These range from the various views of modern day contactees, through the life and events of UFO witnesses and abductees, to the most conservative views of the mainstream scientific community.

A View from the Observatory

One segment of the scientific community in specific has a lot to say about the topic of extraterrestrial life and intelligence. This group is comprised of those involved in SETI, the Search for ExtraTerrestrial Intelligence.[lix] In the SETI community, it is accepted as a given that extraterrestrial life does exist. It is assumed that, when enough facts are known, we will find that intelligent life exists on other planets in the cosmos.

However, in the SETI view of the ET realm, they generally assume that no cosmic visitors have yet come to Earth. At its most conservative, the assumption is that the speed of light is an absolute and unbreakable barrier. The tremendous energy requirements would preclude travel between the stars.[lx]

However, we can assume that, like us, extraterrestrials would be intelligent and curious. Since the most interesting ones would have a level of scientific development at least as high as our own, they would have developed radio or a similar means of communication. Thus, with starflight impossible and electromagnetic communications available, it is assumed that the natural (and presumably the only) way to communicate with other star systems would be via radio.

From the perspective of the SETI community, and based upon our society's scientific knowledge to date, what can we assume about extraterrestrials? The trail of theories and speculation about our potential ET counterparts has already been thoroughly blazed by those in the SETI community.[lxi] Beginning with the early 1960's SETI efforts of Frank Drake and project Ozma, followed by subsequent SETI efforts during the 1970s and 1980s, and more recently with the efforts of the SETI Institute, we have seen ideas of contact mature from a fringe school of thought into an experimental endeavor largely accepted in the scientific community.

This growth has been guided by a faith in the likelihood that there is an ET civilization out there waiting to be heard by us. It is a likelihood determined by a bit of mathematics known as the Drake equation.

A Review of the Drake Equation
The Drake equation is a straightforward formula, developed during the 1960s by astronomer Frank Drake, to determine the probabilities of intelligent life existing somewhere out there. From this equation, knowing the approximate number of stars in the galaxy, one can estimate the number of civilizations which we can contact using a SETI-like endeavor.[lxii]

Ranking highly among the many excellent books on potential contact with ET life, is *Making Contact*, a collection of articles edited by Bill Fawcette.[lxiii] This book provides (in my view) one of the best descriptions of the theories behind the Drake equation. In it, multiple authors discuss the various considerations that make up such an estimate, based upon contemporary knowledge of astronomy, biology, and sociology.

Making Contact extensively describes how estimates determining the potential for intelligent ET life are summarized mathematically in a series of numeric factors. When multiplied together, these factors yield an estimate of the number of civilizations with whom contact is possible.

112

1. R = the rate per year of star formation for stars like our own: Our sun is a type G, undistinguished main sequence star, presumably formed in a non-extraordinary way. How many stars are formed per year in this manner? One estimate is that this rate is approximately one star per year.

2. f_p = the fraction of stars with planets: In "Making Contact" Fawcette develops an estimate, based upon the ongoing discoveries of extrasolar planets, that any sun-like star will most likely have a planetary system of some sort. Thus, he argues that this factor is probably close to 100%.

3. f_e = the fraction of planets that are earthlike: In how many solar systems will there be one or more planets that are like our own world? Fawcette examines computer simulations and current theories of astronomy to develop a conclusion that this fraction is probably somewhere between 50% and 100%.

4. f_l = the fraction of earthlike planets with life: Given the planet's ability to support life, what is the probability that life will occur? Fawcette cites recent findings by geologists that life appears to have been present as far back as can be detected in the fossil record. This suggests that life emerged nearly as soon as the primitive Earth could support it. This suggests that whenever life is possible it will almost certainly come into existence. Thus this number is taken to be 100%.

5. f_i = the fraction of life-bearing planets that develop technological civilization: This is a more speculative factor than those we have previously considered.

   In *Making Contact*, Fawcette covers the many arguments for and against the hypothetical rise of intelligence and technology on a planet. He cites arguments, based upon the concept of parallel evolution, in which similar adaptations evolve to fill similar ecological niches in different locations. Bats, birds and insects all have evolved flight. Dolphins, primates, insects and other animals have all evolved social structures. Many different animals use tools. In any ecological system, life seems to find what works. Thus, he argues that, given enough time for it to evolve; technological intelligence should arise at least once on any given world. Therefore, in *Making Contact* the argument is presented that the rate is 100%.

However, a key question is what it means for a civilization to be technological. In the SETI paradigm, the term "technological" implies that the civilization is radio-capable. For UFO studies of course, "technological" would mean having the capability of interstellar travel, presumably on a routine basis. Thus, rather than being nearly 100%, given the difficulties of travel among the stars, many scientists speculate that this would be extremely rare.

6. L = the effective life of civilization:
   This is probably the most speculative of the factors in the Drake equation. Within this factor lurks every political, environmental and social question that could affect the lifespan of a civilization. Estimates of L range from a few hundred years to tens or hundreds of thousands of years.

7. The result; N = the estimated number of ET civilizations:
   The resulting Drake equation calculates the estimated number of contact-capable ET civilizations by simply multiplying these numbers together to yield the equation:

   $$N = R * fp * fe * fl * fi * L$$

Drake: N = L

From this equation and these estimates, the most optimistic guess is that every to every-other planet-bearing star system would be likely to develop a technological civilization at some point in its history. Thus, according to this model, our interstellar neighborhood should be teaming with life. By using just a little math and multiplying the star formation rate of one per year times the approximately 100% values of the factors yields a result of one civilization per year. Thus, the equation,

$N = R * fp * fe * fl * fi * L$ reduces to $N = L$.

Since L is the lifespan of a civilization and N is the number of civilizations, this implies that the number of civilizations is roughly equal to the number of years in which a typical civilization exists.[lxiv]

Fermi Paradox: N = 1

One objection to the above estimate of $N = L$ is ascribed to the physicist Enrico Fermi when he is reported to have asked, "Where are they?" His objection was that, given the number of resulting civilizations, at least some percentage of them must have been able to develop interstellar travel. He claimed that there have been no visitors to Earth from the stars.

Thus, according to the "Fermi paradox," the number of extraterrestrial civilizations might actually be zero.[lxv] Therefore he suggests that $N = 1$, with the only intelligent race of beings in the cosmos being humanity.

115

Encouraging Factors

While debate rages regarding the various factors in the Drake Equation, there are several principles that offer considerable encouragement that humanity does indeed have neighbors.

1. The Principle of Mediocrity: One guiding principle of the SETI philosophy is that there is nothing special either about Earth or about humanity. What has occurred here on our "pale blue dot"[lxvi] can be assumed to be representative of what could happen anywhere in the universe. Thus, we can assume that since life has evolved here it is probably common elsewhere, as well.

2. Planet hunters are finding it increasingly likely that there are planetary systems around a majority of main sequence stars. At the time of writing of this book, there have been hundreds of planets found around stars within a few hundred light years of our sun.[lxvii]

3. Increasing evidence suggests that life is nearly inevitable given the chemical/physical opportunity to get started. As I write these words the possible discovery is being made that life may exist on Mars,[lxviii] independently of life on Earth. In addition, fossils of ancient microbial life are found in some of Earth's oldest rocks.[lxix] These suggest that life arose rapidly on Earth, once conditions were favorable for its existence. Thus, it appears that whenever conditions are favorable for life, it will arise.

Lasting Civilization Capable of Contact

As we saw above, the "N = L" result suggests that the most important factor determining the number of advanced civilizations in the cosmos may be how long a civilization can survive.[lxx] Estimates of "L" vary widely with a minimum value of 100 years, the approximate number of years in which our present civilization has been capable of emitting radio signals. This low-end estimate assumes that we are in imminent danger of destruction due to nuclear war or environmental catastrophe. It extrapolates the most pessimistic view of our present situation to conclude that many civilizations might enjoy only this same brief flicker of existence.

Higher-end estimates of societal life-span range from tens to hundreds of thousands of years. In the most optimistic estimate, it is theorized that once a civilization lives beyond the current danger period - the advent of modern warfare, nuclear weapons, etc. - it will have achieved the means for its own long-term survival. Thus, perhaps a mature civilization could survive indefinitely.

The historical record of our civilization appears to be on the order 10,000 years (if we consider some of the earliest archaeological sites as indicators of our civilization's beginning). Therefore, as a rough estimate, we can imagine the technological lifespan of a stable but advanced civilization as being at least that long. If we assume that humanity can get through the challenges of the nuclear age and avoid environmental collapse or other disasters, and if our technological civilization could reach an additional 10,000 years of age, it would imply that other civilizations would be able to do the same. Thus, from this viewpoint, a suggested number might be about 10,000 potentially contactable civilizations.

From this discussion, we see that there is little agreement on the value of "L." The only consensus seems to be that we know very little about long-term sociology and prognosis of civilizations, be it our own or anyone else's.

Space Travel and Contact

Arguments in this chapter suggest that there might be many civilizations in the cosmos with advanced technologies. Thus, while the number may be arguable and wildly speculative, future contact seems likely.

In the present day, our world is on the verge of becoming a space faring civilization. With extended duration stays on the International Space Station, long-term spaceflight has now become a reality. Thus, we can assume that interplanetary space travel, at least within our own solar system, is not too far in the future.

If we assume that what we are able to do others can most likely do as well, it becomes ever more plausible to imagine that others might visit, or in the past might have visited, our world from elsewhere. From this view alone, the idea of contact with visitors from elsewhere becomes an increasing consideration. Thus, as we venture ever deeper into space and continue to conduct increasingly sensitive SETI searches, it seems ever more likely that contact will occur.

A Spectrum of Contact

Our speculations about the likelihood of ET civilizations and thus of alien contact, give rise to some of the most intriguing questions in the contemporary world: What would contact be like? What would its impact be on human society? Finally, given the potential impacts, we can ask what contact scenarios the existing social and authority structures on Earth could accept.

Futurist Michael Lindemann[lxxi] has presented an excellent series of ideas regarding contact with ET life.[lxxii] Based upon his model, which I refer to as the "Spectrum" model of contact, we can think of contact as having a range, or spectrum, of "extremeness." Such a model enables us to examine a multitude of questions about ET contact - real or hypothetical - in the past, present, or future.

At the "low end" of the spectrum are the least impacting forms of interaction, such as discovery of extraterrestrial microbes or fossils from ancient primitive life on other worlds. At the highest end of the spectrum would be the maximum-impact contact scenarios such as overt ET arrival, the proverbial landing on the Whitehouse lawn.

In his model, Lindemann describes a set of factors, or variables of contact, which would govern the nature of an interaction with ET:

- Speed: What is the time scale over which contact occurs - therefore, how much time is available for humanity to adjust to the contact scenario?
- Manageability: How well can human society manage/control the contact scenario?
- Strangeness or Impact: How much new information is infused into human society, challenging or stretching (threatening?) existing beliefs? How far do the paradigms introduced by contact diverge from current human thought?

Range of Contact Scenarios

Lindemann offers a series of examples ranging from the least impacting to the most impacting:

1. Ancient or Existing Primitive life:
    *   Signs and Signals: Discovery of Fossil Evidence - (ET life in the distant past: the ALH84001 meteorite found in Antarctica, apparently Martian in origin),
    *   Presently existing simple life on Mars or Jupiter's icy moon Europa,

2. Intelligent Life:
    *   ET Artifact Discovery: Indicating that Civilization existed nearby, but a long time ago, such as the alleged Face on Mars,
    *   A bona-fide SETI detection: Civilization out there in the present day, but very far away,
    *   Historical Contact: Verifiable discovery of ET contact in the archaeological or historical record, suggesting that "they" came here a long time ago,

3. Arrival:
    *   UFO Sightings: Covert presence, observing us in the present day, not including possible contact with UFO occupants,
    *   Abductions and covert contact: This includes the currently understood CE4 scenario, indicating that the visitors are, in some way, in covert interaction with us in present times.
    *   Overt contact: This is the ultimate scenario of overt arrival, in which ETs arrive and reveal themselves to us. It is the landing on the Whitehouse lawn.

Using the model we have just described, let's consider each of these scenarios in turn, and draw our own conclusion regarding the variables of contact.

Ancient or Existing Primitive Life

The last few years have seen an increasing acceptance of the possibility of extraterrestrial biology. The search for possible primitive life on other worlds has gained a strong following in the scientific community. The NASA Origins program[lxxiii] has initiated a formal research effort into the possibility of life on other planets. Furthermore, the search for possible life (past or present) on Mars[lxxiv] and Europa[lxxv] has now become a key focus in the exploration of our own solar system.

Initially, Lindemann's model considers the simplest, least-impacting case of "contact," the discovery of simple, non-intelligent life on another world. This would include the discovery of extraterrestrial microorganisms, or fossil evidence of ancient life.

Lindemann offers the following as the first two scenarios.

Signs and Signals

Lindemann first considers the scenario in which we discover possible signs of ancient life. Here, he states:

"A fossil in a rock is a sign - not life itself, but a strong indication that life once existed."

The first such tantalizing hints of evidence for extraterrestrial life occurred with the discovery of possible microfossils in the meteorite fragment, dubbed ALH84001.[lxxvi] Most scientists believe that this meteorite originated on the surface of Mars in the distant geologic past. In 1996, scientists discovered evidence of possible ancient biology within this rock fragment.

The possibility of fossil evidence of life in rock from another world opens the door to the very real possibility of extraterrestrial life. Yet, while such extraterrestrial biology is acceptable in principle, many in the scientific community have since greeted with skepticism, the claims of fossilized life in the ALH84001. It is not clear now whether these micro-formations are fossils and the result is in doubt today. However, in provoking vigorous debate, this discovery has opened the door to consideration of such possibilities.

Lindemann states that for this scenario there would be:

- Low to moderate speed of impact,
- Moderate to high manageability,
- Moderate strangeness.

While interesting and of high scientific value, there is little overall challenge to the existing scientific views of our world. In summary, Lindemann states:

> *"In general, this kind of contact scenario would be the most desirable in terms of maintaining social order."*

Present Microbial ET Life

The next step, beyond the discovery of fossilized ancient life would be the discovery of presently existing, but biologically simple life. Regarding this scenario, Lindemann states:

> *"In the minds of many space scientists, it is becoming more and more probable that we will soon discover living extraterrestrial organisms, perhaps on Mars or Jupiter's water moon Europa. Just as most of life on Earth is relatively simple and non-intelligent, so too we should expect that a vast menagerie of "lower" life-forms might inhabit the limitless expanse of space."*

As this quote suggests, an intriguing scenario is the possibility of life on Europa, a moon of Jupiter.[lxxvii] The most recent space probes to Jupiter have indicated that Europa is probably an ice-covered world with an ocean of liquid water beneath. It is a tenet of biology that where there is water there can be life. Thus, the possibility exists that there may be some form of life, presumably primitive, in the hidden seas of Europa.

In addition, much of the current exploration of Mars focuses upon the possibility of finding primitive life. Thus, the possibility of finding life on one of these worlds is significant.

The discovery of present-day extraterrestrial life would be a fascinating event.
Here Lindemann states:

"Finding life anywhere in the universe beyond the Earth would be among the most momentous discoveries in history. But microbes on Mars, while scientifically enthralling, would pose little threat to Earth as long as they were handled with the care accorded any potential biohazard. Nor would they be seen by most people as promising sudden advancement or other great change in the human condition."

In terms of the variables of contact, Lindemann suggests that, for these cases, there would be:

- Moderate assimilation speed
- Moderate to high manageability
- Moderate strangeness

As we have seen above, it is potentially straightforward to handle these concepts within existing social/scientific paradigms. Here, Lindemann states that:

"The discovery of non-intelligent ET life at a distance - even as close as Mars - might be only moderately impactful in the long run."

While exobiology, the study of extraterrestrial life, can bring advancement at the leading edges of biological science, it is also within the existing paradigms of science: Darwinian evolution and contemporary physics would continue to apply. Thus, while pushing the leading edge of science, this area of study remains acceptable to conservative scientific institutions.

In addition, the discovery of extraterrestrial microbes would not present humanity with any external timeline. It would pose no threat to existing human institutions and no challenges to prevailing social order. Thus, we humans can take this entire endeavor at our own rate.

Would such ideas have been acceptable even 20 years ago? The mere fact that this is now a recognized possibility within the halls of mainstream science means that the baseline of scientific thought has stretched irrevocably outward to accept the existence of ET life. Thus, the prevailing social paradigms, even in our most conservative scientific institutions, are evolving to allow the search for extraterrestrial life, albeit in its most primitive form.

The next generation of planetary exploration and exobiology should prove to be both fascinating and profound. In our own solar system, Mars and Europa might just become real-life laboratories for ET biology. At the same time, the NASA Origins program will search for signs of biospheres existing on extrasolar planets. While not pushing the most conservative institutions of humanity too far beyond their comfort zones, these projects will continue to stretch the thought envelope. Slowly but surely the concept of extraterrestrial contact is becoming acceptable.

Intelligent Life

We now move to the next level of significance. Up to this point, we have allowed for the possibility that life might exist, albeit in its most primitive form. We now ask what if such beings were intelligent, with an intellect equal or greater than our own. Suddenly, we are no longer alone in the universe. Now, we find that, in some way, we must share our universe with others.

Exo-Artifact Discovery

Let's first consider the potential discovery of exo-archaeological artifacts on another world. We can use as an example, the alleged Face on Mars.[20] If this were to turn out to be valid discovery, what would the reaction be in human society?
In this consideration, Lindemann states:

"...it is possible that humans could discover a far more dramatic sign of ET life: an obviously intelligent artifact. For example, if the so-called "Face on Mars" were proven to be an artificial structure, the implications would be quite earth-shaking, even though no direct contact with ET life had occurred."

Lindemann further suggests the possibility that the creators of the face on Mars, if it is an artifact, would not be native to that planet. He states:

"If an extraterrestrial intelligence did leave monuments on Mars, could such an intelligence have evolved there? Our current understanding of Mars suggests that such a scenario is unlikely. Where then, could such an intelligence have come from?"

---

[20] Note: It is acknowledged that there is considerable skepticism about the "Face on Mars." In this author's view, the skepticism may be warranted. However, for the purposes of this analysis, let us imagine that it is indeed, a bona-fide phenomenon of artificial, extraterrestrial origin.

Such a discovery would imply that at some point, perhaps a very long time ago, at least one civilization existed on Mars. Let us assume, as in the above comments, that the inhabitants were not native to Mars, but colonized Mars from elsewhere. This would directly imply that at some time in history these beings had developed the capability for interstellar travel. This would be required to get to Mars from an origin, which presumably would be in another star system. This alone, would conclusively establish that star-faring intelligent beings have existed - and thus can exist - in the universe.

The face on Mars or other similar discovery would spark tremendous curiosity, yet have a relatively low impact on society. Thus, although Lindemann does not offer specific variables of contact for this scenario, here we can extend his model to suggest that there would be:

- Still moderate impact: Such a discovery would have very little direct impact on our society. Yet it would force many scientific questions: Where are they now? What were/are they like? Moreover, what happened to them?
- Still moderate to high manageability: These beings would not be any threat to us, as their existence would be safely in the distant past.
- Moderate speed: The phenomenon we have discovered is passive and we could proceed at our own rate to explore this mystery.

Such a discovery would be a paradigm-challenging event. However, it would very likely not cause irreparable harm to humanity and would, in the end, result in a tremendous expansion of the human knowledge envelope.

SETI Detection: ET far away, in the present time
Let's now move from the idea of life being safely in the past or comfortably primitive, to consider the discovery of contemporary intelligent ET life. What if SETI were to detect a clearly intelligent radio signal from another world? What effect would this have on human individuals or institutions, scientific or otherwise?
In this regard, Lindemann states:
> *"There can be no reasonable doubt that such a discovery would have huge and lasting impacts upon many aspects of human life. Among the current generation of astronomers and space scientists, many still doubt that extraterrestrial life possessing human-like intelligence, or greater, exists elsewhere in our galaxy."*

As Lindemann states, a verifiable SETI detection would have considerable social and scientific impact: it would establish that intelligent extraterrestrial beings exist now![21] Indeed, it would challenge the deeply held views of some of the more conservative members of our society, scientific and otherwise, that humanity is alone in the universe.
Furthermore, this would pose a number of theoretical challenges. Foremost among them would be to understand the alien message. Since a message must be transmitted in some type of language or code, how would we decode it to determine what they are saying? Moreover, if we choose to respond, how would we begin to talk with them?
Regarding any potential human response to an ET signal, Lindemann states:
> *"Because the distance is likely to be ten light years or more, our response will take at least ten years to arrive. It would not make sense, therefore, to send only short bursts*

---

[21] Strictly speaking detection of an artificial radio signal from an interstellar source would imply only that a technological civilization existed at the time in which the message was transmitted. For a distant signal source, a significant duration could have elapsed since the signal was first transmitted.

and wait twenty or more years for each response. Instead, what might happen is that both sides will soon begin sending steady streams of communication."

He further states:

"Here on Earth, coordination and regulation of the SETI response effort could quickly become a policy nightmare. Who speaks for humankind? Who controls and disseminates incoming information? How can pirate broadcasts, reflecting different slants on interstellar diplomacy, be averted? How might signal hoarding change the global balance of power?"

As this quote suggests, the possible political and sociological ramifications of a radio-SETI dialog with a distant civilization would be far-reaching.

In terms of the effect upon our society, we can apply Lindemann's variables of contact, speculating that a SETI detection would have:

- Moderate to high manageability: A SETI detection would raise many philosophical questions, but as the ETs are still comfortably distant, the life of human society would continue relatively undisturbed. Though, as Lindemann suggests, a SETI signal might have significant political impact.

- Moderate strangeness: At least in principle, we could understand the medium, the physics and potentially the language of contact.

- Moderate speed: A SETI signal would probably be of short duration, but again, we would have plenty of time to analyze the message and determine what (if any) response humanity should make to it.

The above considerations are of the detection of a verifiable ET signal. Yet, even if no detection ever occurs, we find that the search itself has fostered a paradigm shift within western society.

In the recent past, even as recently as 15 years ago, the topic of a search for extraterrestrial intelligence, even at its most conservative, was still ridiculed in political officialdom. This was graphically demonstrated by the demise of the NASA SETI project, the High Resolution Microwave Survey,[22] killed by the stroke of the congressional budget axe when ridiculed as a pork-barrel project "chasing space aliens."

Yet, the result was that the SETI effort continued, not by government, but by farsighted individuals within the private sector. The vision is present, though not all can see it. Now, because of these efforts, the paradigm has advanced to the point where the idea of searching for (and contacting?) extraterrestrial intelligence has become acceptable.

Historical Contact: ET visiting Earth, long ago

Another scenario of intermediate impact would be that of contact between humanity and extraterrestrials, at some time during human (pre)history. What would the impact be if we were to learn conclusively that our history had been directly influenced by contact with ET visitors?

There is indeed suggestion that such a scenario may have occurred. Carl Sagan wrote about such possibilities in his book, *The Cosmic Connection*.[lxxviii] In it, Sagan cites examples of possible historical anomalies such as the religious rituals of the Dogon tribe in Africa. As part of their religion, they ritually re-enact a phenomenon that is not visible to the naked eye, the transit (or passage) of one star, Sirius B, in front of its companion star, Sirius A. Astronomers discovered this invisible companion to Sirius in 1862, thousands of years after the beginnings of the Dogon ritual. Sagan asks the question of whether this tradition could have had an extraterrestrial origin. In addition, a number of additional works[lxxix]

---

[22] The High Resolution Microwave Survey was a NASA project, active during 1992 and 1993, with the goal of detecting possible signals of intelligent origin from interstellar space. The project is described in online articles at the following locations:
http://www.aas.org/publications/baas/v25n2/aas182/abshtml/S7101.html and
http://www.totse.com/en/technology/space_astronomy_nasa/nasasrch.html

explore, in depth, the possibility that Extraterrestrial beings influenced the Dogon ritual.

Similarly, the abduction researcher, John Carpenter, in his paper, "Centuries of Contact,"[lxxx] suggests that there may have been extraterrestrial influences on humanity throughout history. Carpenter cites possible UFO appearances in ancient art, historical accounts and folklore to develop a theme of ancient ET contact.

Another excellent work in the historical-UFO/contact literature is Dr. Barry Downing's book, *The Bible and Flying Saucers*,[lxxxi] which examines the Bible in terms of possible descriptions of UFO and/or alien contact. Several other works, including Raymond Fowler's book, *The Watchers*,[lxxxii] also suggest a scenario in which ET beings have been interacting with humanity throughout history. Historical researcher Zachariah Sitchin[lxxxiii] has developed this theme in considerable detail, arguing that humanity is the direct result of influences by extraterrestrial beings.

What if the archaeological community should find unequivocal evidence that extraterrestrial visitors had visited us in the distant past? We could indeed view this as an intermediate scenario, in which the impact would be relatively high on certain portions of the scientific and religious communities. We can once again apply Lindemann's model to speculate on the following variables of contact:

- Impact: Low to intermediate for most of humanity. We would continue to lead normal lives. However, the impact would be high for the scientific and religious community, as we would need to revise our views of history significantly.
- Manageability: Intermediate. It would be difficult to control the effect that knowledge of ancient contact would have on the population, particularly within the scientific and religious communities. Yet this would be mitigated by the sense that, while it might affect our views of our world and our history, there would not be any immediate change to our daily lives.
- Speed: Still low. The phenomenon would still be passive and would be strictly informational in nature. Its overall impact on our society would probably be relatively slight.

ET Arrival

The ultimate scenario in Lindemann's model is that of alien arrival on our world. Instead of distant or ancient contact events, we must now consider the idea that extraterrestrial entities might somehow be in contemporary contact with us.

In such a scenario, humanity would have little control over the interaction. However, Lindemann's model suggests multiple degrees, a range of scenarios from covert observation to overt arrival. Here, we examine this range and the corresponding impact of each scenario.

Watchers: UFOs observing us covertly

The least impacting of these scenarios is that of the contemporary UFO sighting. This suggests that the visitors are observing us now, though not interacting with us in any direct way.

A bona-fide UFO reality directly implies the presence of alien beings in our vicinity. Thus, even though they might not directly interact with humanity, remaining as "Watchers" in the distance, we know that they are present, therefore uncontrolled and potentially unpredictable. They are beyond the authority and perhaps the understanding of human governments. Thus, the impact of a UFO reality would be considerably higher than earlier scenarios.

Lindemann states:

"If the Watchers scenario is real, one can only conclude that it is far gentler -- perhaps intentionally so -- than that of a frontal arrival. Even considered theoretically, the Watchers impact is relatively subdued"

Depending upon the extent of the UFO phenomenon, Lindemann suggests variables of contact of a covert UFO presence in our skies:

- Strangeness: Moderate to high

- Manageability: Moderate to high

- Speed: Moderate

As the above indicates, the UFO phenomenon has the potential to affect our worldview tremendously. Any initiative taken by the phenomenon would leave little time for humanity to react. UFO behavior is outside of human understanding and all interactions are under the control of the visitors.

Yet as we see, even with the apparent presence of the alien beings as possible agents behind the UFO phenomenon, governments on Earth appear to function today. The phenomenon has thus far remained covert, minimizing the need for rapid reaction by human society. Human activity continues unaffected - at least for the short term.

Abduction: UFOs/Visitors interacting with us covertly
We now come full-circle, returning to the mystery of close encounters and our relationship with the phenomenon. In earlier chapters, we discussed the effect on the individual experiencer. We also saw the massive scale of the UFO abduction phenomenon. While the effect on the individual is powerful, ultimately we must consider the UFO abduction phenomenon in aggregate.

What is the visitor agenda? How does covert contact on such a large scale affect society? We can adapt Lindemann's variables of contact and suggest that such covert-interaction may have:

- Intermediate to high impact: The UFO abduction event is one of very high strangeness. As more than one UFO researcher has suggested, the one thing we can be sure of about aliens is that their ways and their thoughts will be alien to us. Alien behaviors and motivations seem to defy our understanding. Thus, the scenario of present-day-visitation, even though covert, can have powerful ramifications for our present civilization.

- Low to moderate manageability: The phenomenon is entirely outside of human control and defies any efforts of human agencies to control it. Yet, as long as the phenomenon remains covert, human civilization and authority will continue to function more or less intact.

- Intermediate speed: This is still moderate, but under control of the visitors. As we stated above, the phenomenon appears to remain covert, preferring to interact with humanity at the individual level, in the dark of night. For better or for worse, the visitors, themselves, seem to be limiting the rate at which humanity must adapt. Yet while the UFO phenomenon remains covert, we saw in earlier chapters how the effect upon the individual experiencer is immediate and profound. Contact affects the experiencer right now!

Maximum Impact - Overt Contact

We now examine the most extreme case, that of unexpected, overt contact. What would be the impact of such an absolute maximum scenario? How would humanity react?

We can imagine the fabled "Whitehouse Lawn" Scenario, in which the aliens, whoever they might be, land at the seat of government (the 1952 movie, "The Day the Earth Stood Still"), or hover over our major cities (1996 movie, "Independence Day"). In such cases, Lindemann states:

> *"In all likelihood, whether arriving ETs were actually benevolent of malevolent, human fears would be aroused; human reactions would be highly impulsive and might range from swooning adoration to extreme violence. Such reactions were seen during the approach of Comet Hale-Bopp in early 1997. Some people believed the comet was accompanied by a companion spacecraft, a tragic misperception that led to the suicides of at least 41 members of the Heaven's Gate cult in southern California."*

As Lindemann suggests, the social ramifications of such contact would be tremendous indeed. The summary of "The Implications of a Discovery of Extraterrestrial Life" within the 1960 Brookings Report to NASA[lxxxiv] describes significant potential negative reactions to extraterrestrial contact by (as a minimum) the scientific and religious communities. Depending upon the nature of the contact, it could disrupt large segments of society in unpredictable ways.

Hale-Bopp and Heaven's Gate

As suggested above, we can think of the March 1997 arrival of the comet Hale-Bopp as a dry run of an overt contact. At the time of its closest approach, there were reports in the new age media that there was an alleged "companion" object (e.g. a UFO). The Heavens Gate cult took this as a signal that they were to return to the heavens, presumably in non-physical form. To do so, they believed that they needed to "shed their containers." Thus began the path to suicide for the 39 members of Heaven's gate.

Budd Hopkins, in his talk at the 1997 MUFON symposium,[lxxxv] asked the question: "What if The Hale-Bopp companion had been real?" In his talk, he indicated (correctly, I think) that this incident had significant implications for those contemplating overt contact. If a comet flyby resulted in scores of people committing suicide, what would the ramifications be if the Hale-Bopp comet approach really **HAD** involved a UFO? To Hopkins, this suggests that "enlightened" ETs would most likely postpone overt contact.

From the above discussion, we can speculate that there would potentially be significant negative impact from direct, overt contact, especially if it occurred on a large scale. Such a scenario would probably illustrate the extreme high end of the variables of contact. Lindemann states that the variables of contact would be:

- Strangeness or Impact: Very high,
- Manageability: Low,
- Speed: Rapid.

We can imagine that, in the event of a "Whitehouse lawn" or "Independence Day" scenario, there would be very little time for humanity to react. An overt large-scale contact would have a minimum of manageability and would result in absolute maximum impact. Their motives would be alien to our concepts and would probably involve an absolute maximum of strangeness. I can picture no part of life unaffected by such an event.

Furthermore, the contact process would be entirely under control of the ETs. We would probably have little understanding of their capabilities (military or otherwise). Thus, our well-being and perhaps our very survival would be dependent upon their intentions, be they friendly, indifferent or hostile.

Cultural Contacts in History

We can get a less-than-positive glimpse of the ramifications of massive overt alien contact by studying contacts throughout history between human civilizations. Historical precedent seems to suggest that lower technology civilizations do not fare well in contact with those of higher technology. The history of North America contains graphic examples of how such contact with the higher technology European civilization largely destroyed the lower technology indigenous civilizations. This alone provides a case for potential negative effects of contact.

Who speaks for Earth?

Another problem with overt contact: Who on our world would extraterrestrials actually contact? When we refer to a "Whitehouse lawn" scenario, whose Whitehouse lawn would they land on? Would it be that of a particular nation, perhaps the United Nations? Would they select noted scientists? Or would the visitors choose to contact certain abductees/contactees? In short, who speaks for Earth?

At present, there seems to be little or no agreement on this among scholars of the topic. Thus, at the present time, at least within the SETI community, if contact does occur, the SETI Declaration of principles[lxxxvi] forbids response until this question is resolved. However, this agreement is non-binding and we have seen several private efforts to engage in two-way contact.[23]

The biggest example of a private contact effort is the attempt by the Center for the Study of ExtraTerrestrial Intelligence (CSETI) to interact with our presumed ET visitors. Dr. Steven Greer, head of the CSETI organization, describes the CSETI CE5 Initiative as "an attempt to vector UFOs into a specific area for the purposes of initiating communication."[lxxxvii]

---

[23] We can think of passive human-originated artifacts such as the Voyager and Pioneer space probes, as indirect or unintentional contact attempts. Each such probe bears a message from humans to any extraterrestrial beings that might recover it in the (presumably very distant) future. However, these are haphazard and very indirect "contact attempts" and so cannot be considered real time contact in any sense.

It is likely that, in the coming years, more such private efforts will arise. The effect of these, upon individuals involved and on society overall, cannot fail to be both fascinating and unpredictable.

Contact and Popular Mechanics

In the February 2004 issue of Popular Mechanics, an article appeared entitled "When UFOs Arrive."[lxxxviii] In it they discuss the possibility of first contact (which they clearly state is still in the future) and what the presumed human reaction would be.

The article considers two main classes of reactions, one from the scientific community and the other from the US government/military. The article initially describes how there will be "obvious signs of their approach," presumably radar or radio contact. It then describes the SETI Declaration of Principles of first contact, which we could roughly describe as "be absolutely sure and then tell everybody."

This reaction-scenario assumes that the scientific (e.g. SETI) community would somehow be the leaders in a contact scenario. The article then goes on to demolish this assumption, describing what would most likely **actually** happen, a government and military takeover of the contact scenario. This would include a heavy-handed treatment of the visitors by human authorities.

We can glean a hint of this from the Brookings Report of 1960, which describes many of the implications of open first contact as envisioned at that time. It suggests that contact would be harmful to humanity and recommends government responses to ET arrival. The resulting government policy appears to be one of heavy-handed hostility towards any ET visitors unlucky enough to fall into the hands of human (at least U.S.) officialdom. In the section entitled "State of Emergency," the article reads as follows:

> *"If ET Turns up at NASA's doorstep, bearing that invitation [CL Note: invitation from the plaque on Pioneer 11], it is in for a surprise. Instead of getting a handshake from the head of NASA, it will be handcuffed by FBI agents dressed in Biosafety Level 4 suits. Instead of sleeping in the Lincoln Bedroom at the Whitehouse, the alien will be whisked away to the Department of Agriculture's Animal Disease center..."*

138

Could this be an actual description of US policy in the event of contact with ET visitors? A suggestion of such events appears to have occurred following the crash of an apparent extraterrestrial object in 1965 at Kecksburg PA.[lxxxix] According to witness descriptions, in a very short time the military had thoroughly secured the scene and had recovered the object, which was never again seen by anyone without a high security clearance. This suggests that the mainstream scientific community would probably be locked out by the military powers that be. Thus, they would most likely find themselves subject to the same apparent coverup that has bedeviled UFO research for the last 50-plus years.

So what does this suggest would be the most likely contact scenario? Would ET simply walk into a situation such as this? Presumably, they would be well aware of the situation on any world they were contacting (e.g. us). Thus, they would presumably not walk into the ugly situation described in this article. We can assume that they would choose some means of contact that would be under their control and on their terms. This leads us to the question, is such contact already occurring?

The Gradual, Covert Contact Process

The potentially severe problems associated with open contact such as a "landing on the Whitehouse lawn" would imply that in the foreseeable future the visitors would most likely avoid it. In order to minimize their impact, they would probably employ at most, a gradual, controlled release of information about themselves. We can further imagine that an ET civilization would initially employ covert method(s) to acclimate us gradually to their presence. We can imagine a gradual process of acclimation of humanity to a more advanced ET society, long before any kind of open contact occurs.

Is this what the visitors are presently doing? In the next chapter, we will look at potential strategies an extraterrestrial society might use to initiate contact with Earth and how this might appear to humanity. We will then consider whether this might indeed be the strategy currently employed by visitors to our world from across the Cosmic Bridge.

Chapter 8 – Strategies of Contact, a Phenomenal Agenda

Picture in your imagination a cosmic civilization, a vast spectrum of different species and cultures. Imagine that it is governed in some enlightened way, maintaining order for many thousands of years - an evolved, ancient, stable federation of worlds. Now imagine that at the periphery of this realm is a new world, one with an up-and-coming civilization. On this distant blue planet, the beings are both civilized and warlike, poets and killers, saints and thieves. They are diverse and quarrelsome but they hold promise. They might bring a new creative spark to the cosmic realm but they might also bring danger.

Might such a scenario exist? Could there be such a vast cosmic civilization? Might we be pending newcomers to such a community of which we presently know nothing? Might there, in fact, be such a super-civilization keeping an eye on us, ultimately preparing us for contact?

In the last chapter, we examined some classical concepts of extraterrestrial contact and suggested that the result of overt contact between human and cosmic societies would likely be negative. Thus, we can ask how (or even if) visitors from elsewhere might safely seek to open the door to interaction with humanity. How could they interact with humanity without severely damaging our society?

To explore this question, in this chapter we will examine some possible strategies which an ET civilization might employ to close the gap between them and us - to build the Cosmic Bridge. We will speculate on whether the visitors, who or whatever they may be, might actually be initiating contact. In addition, we will suggest some possible reasons they might have for doing so.

Are They Hostile?

Often, one of the first questions I hear in discussions of alien contact is "are they friend or foe?" My answer is always, "I don't know." Yet the history of our interaction with the UFO phenomenon shows that they almost never harm humans. There have been many encounters between the visitors and the various militaries of Earth.[xc] In each case, the visitors technology shows clear superiority over that of the human military. UFOs can come and go at will, with little interference from humans. Yet we have not experienced direct military conquest by their alien occupants.

Given the technological superiority demonstrated by UFOs, I believe they would easily defeat human military forces in any direct confrontation. The fact that they have not done so and apparently try to avoid confrontation, suggests that they are not hostile. While their intentions are ambiguous, they do not appear to have military designs on Earth. Yet their activities are extensive, a massive agenda with unknown goals. What are they doing here - and why?

A "Possible ET Strategy for Earth"

Could the visitors be conducting a program of orchestrated contact? If so, then what might be their overall strategy? Could the close encounter phenomenon be one stage in such a plan - one in which contact is conducted one individual at a time?

Let's begin asking these questions by examining some of the big-picture models of how the visitors might conduct a contact process, an ET strategy for Earth. Let us assume that warnings of the danger to human society from open contact with extraterrestrial beings are accurate. How could an ET civilization initiate contact, yet avoid the accompanying perils?

In his article, "A Possible Extraterrestrial Strategy for Earth,"[xci] Dr. James Deardorff postulates just such a model of contact, a slow, gradual progression of contact in carefully managed steps.

Deardorff begins with one of the biggest questions posed and perhaps the biggest problem with the traditional contact paradigm. This is the lack of confirmed ET intelligence (ETI) detections to date. According to several possible interpretations of the Drake equation (see chapter 7), there should be a significant number of civilizations detectable to our radio ears. Thus, according to the SETI paradigm, we should have heard something by now. Furthermore, the reasoning goes that if "they" have space travel capability, they should be visiting our skies now and thus should be visible to us. In short, in the words of the physicist, Enrico Fermi, "Where are they?"[xcii]

As we noted in the previous chapter, probabilistic arguments alone suggest that a large number of worlds in our region of the galaxy could be home to intelligent beings. On at least some of these worlds, the inhabitants should have developed a technological civilization - and eventually, radio and/or space travel. Thus, our model of intelligent life in the universe must explain this apparent lack of contact to date. The Deardorff model suggests a way to explain this apparent silence. Yet at the same time, it hypothesizes a large-scale unfolding of covert contact between humanity and the visitors.

Deardorff envisions a staged program of quarantine, observation and gradual contact. Finally, once all previous stages are successful and the time is right, there would be open disclosure of the ET presence and overt interaction between civilizations.

Assumptions

Deardorff and others[xciii] make some very broad assumptions about the nature of potential visitors to Earth. They assume that any world that has survived long enough and become sufficiently advanced to be welcomed into the cosmic community, can be assumed to have survived its potentially warlike youth. We can reason that they will have reached a level of civilization qualifying them for membership in such a cosmic community. Thus, we can probably assume that, at least to some extent, the prevailing ET civilization(s) would be benevolent. Deardorff suggests that we can therefore assume the overall ET civilization is motivated to protect the interests of the developing worlds within its sphere of influence. From this, he reasons that the overall community of intelligent beings would be reasonably nurturing to "up and coming" worlds such as ours.

Contact Strategy

As mentioned earlier, a quick perusal of first-contact speculation, in which the consequences of overt contact are considered,[xciv][xcv] strongly suggests that at our present level of civilization the results would be negative. Thus, Deardorff proposes that at the present time any enlightened ET civilization would avoid open contact. Instead, he suggests that an advanced civilization would follow a gradual plan for the acclimation of humanity to an ET presence.

Deardorff proposes that Earth (and other emerging worlds) would be under some form of quarantine, a Star-Trek-like "Prime Directive." This would serve both to insulate "primitive" worlds such as ours from external influence and to prevent the expansion

of humanity into the cosmos while still in our warlike adolescence. With this quarantine in place, a careful program of contact and acclimation could occur.
There would likely be several stages to such a program:

1. An initial observing stage: mapping, surveying and scientific observation - effectively concealed from human perception but including the close study of humanity.

2. Initial covert interaction: this would include abduction and expanding clandestine contacts.

3. Gradual disclosure of the ET presence: this would occur in subtle ways, beginning at the grass-roots level, but perhaps also secretly at official levels.

4. A Subtle infusion of knowledge of ET topics into the mainstream of human culture: this would include concepts and/or discoveries that shift human paradigms toward the acceptance of contact.

5. As humans eventually became more accepting of an ET presence, a careful but ever increasing level of overt contact could occur.

An evolving contact agenda would most likely begin with an extended period of observation in which the extraterrestrials would study humanity extensively, learning as much detail as possible about our society, technology, psychology, etc. A program of limited contact would follow this, occurring covertly. This interaction would be kept out of the attention of the authorities (political, military, scientific, etc).
Deardorff also suggests that there might be an air of mystery and logical absurdity to the ET contact effort, at least partially due to their advanced technology, "indistinguishable from magic."[xcvi] He further suggests that the ET presence might even

deliberately conceal itself with a veneer of absurdity. One purpose of such deliberately illogical behavior might be to make it seem nonsensical and thus "unscientific," making it uninteresting to the scientific community. In this way, the contact effort would avoid a premature acceptance by the authority structures of Earth - until humanity had evolved sufficiently for open contact.

In the Deardorff model, contact would occur to selected individuals in a covert manner. This process would occur at a grass roots level. It would begin selectively, gradually widening in scope, slowly acclimating greater numbers of people to the ET presence. This furtive but ubiquitous process of contact would be compatible with an overall "Prime Directive" style embargo and would bring about a relatively painless preparation of humanity for contact.

Within the UFO encounter phenomenon, we see many potential signs of such a large-scale acclimation process. Indeed, when one researches the UFO and close encounter phenomenon to any degree one begins to conclude that it is built around this very concept. Nearly every experiencer describes the phenomenon as being covert, coming to him or her in the dark of night.

The vast scope of the abduction phenomenon,[xcvii] which we described in earlier chapters, seems entirely compatible with the bottom-up strategy described in the Deardorff model. The Roper poll and other estimates suggest that phenomenal contact in some form has affected nearly one percent of humanity. We could perhaps view this as an indication of a massive effort to introduce humanity to the visitors' presence.

The phenomenon itself seems cloaked in mystery and absurdity, even an aura of magic. It presents us with a host of logical paradoxes and superpositions of reality, an overlay of dreams and the physical. Is this "magic" simply our (mis)perception of a vastly superior technology? Or is there a deliberate presentation to us of absurdity, in order to repel the "rational" scientific community?

The Deardorff model corresponds closely with some very basic assumptions I would make about humanity:

1. Humanity is not yet ready for contact: As we have suggested earlier, the reactions of humanity to overt contact would not be positive. Thus, the visitors would probably avoid overt contact at the present time.

2. The primary human occupation is war: We have perfected the art of killing each other to a high degree. It is not too difficult to imagine humans treating visitors in this same way. We can only assume that the visitors are well aware of this and thus see us as dangerous.

These factors suggest that the Deardorff model of covert acclimation could be a key to understanding any possible alien agenda. Perhaps the visitors are gradually opening humanity up to contact with the greater cosmic community. We can imagine a goal of eventual open disclosure, when the time is right, presumably culminating in the entry of humanity into what UFO researcher and author, Stanton Friedman has referred to as "The Cosmic Kindergarten."[xcviii]

The Zoo Hypothesis and the "Nature Preserve" Metaphor

In researching this book, I spent a considerable amount of time reading about the Fermi Paradox and its possible solutions. As we described in the previous chapter, the Fermi Paradox asks why we don't see the aliens, given the conclusion that they must be visiting us as the numbers calculated by the Drake Equation suggest.

One potential solution to the Fermi Paradox is a theory called "the Zoo Hypothesis."[xcix] This theory argues that extraterrestrials are indeed visiting us. However, their primary activity is covert observation, monitoring us without being seen. They would seldom leave evidence of their presence, unless such artifacts were somehow important to their plan. Only in the most extreme cases would they make their presence known.

What do such metaphors suggest about our relationship with the visitors? Can we apply such a concept to the abduction phenomenon?

One possible way to apply the Zoo Hypothesis is to imagine Earth not just as a zoo, but as a managed preserve. I call this the "National Park" or "Nature Preserve" (NP) Metaphor. In this analogy, we can think of our world as a "wilderness preserve," carefully monitored by ET "rangers." As in nature preserves set aside by human society, the purpose of this reserve would be scientific research as well as preservation of the environment and the species residing there. To the residents of the preserve the presence of the Watchers[c] would be largely transparent. The animals perceive what appears to be a natural environment.

In this metaphor, in a manner similar to the animals in a nature preserve, terrestrial life would be managed and studied, yet essentially remain "wild." They would have little idea of what actually goes on behind the scenes, what the ET "rangers" are actually doing.

In "wild" parks such as the national forests of the USA, naturalists flying overhead covertly and routinely observe the animals. However, in any wildlife management program, a certain amount of careful, covert "contact" occurs for scientific study and population management. On occasion, animals are captured, studied, banded and released. This gives the naturalists ongoing data about the health and status of the animal population. Yet the process leaves the animals unharmed, only briefly interfering with their lives.

Yet when such "contact" with rangers does occur, we can imagine that to the animal under study, it is a very strange experience. Two-legged beings in flying machines have little place in the world of the wolf or elk. Such encounters probably involve tremendous levels of fear on the part of the animal, which has been rendered helpless and subjected to an experience of which it has little or no understanding.

In this metaphor, we imagine our ET visitors in the role of the naturalists. Frequently, just like human rangers, they fly in, pick up one or more human "animals" and do their scientific activities. They then release the humans, confused but generally no worse for the wear. To the human residents of the Earth-Nature-Preserve, this intervention would appear as the phenomenon known as UFO abduction.

Other aspects of the naturalist/NP metaphor may apply to the abduction phenomenon as well. Whereas park rangers or wildlife biologists attach transmitters to track animals by radio, we also have reports of abductees receiving implants.[ci] Abductees are often under the impression that these are devices for monitoring their status and whereabouts.

Abductees also describe how the phenomenon seems to follow family lineages.[cii] Perhaps this is analogous to the biologist who monitors successive generations of wolves for wildlife management purposes. This might be to track evolution, fertility, behavior, and/or survival trends, a multigenerational research endeavor. Perhaps we are being studied, or even influenced, in an analogous way to the elk, wolves and bears of our own parks.

Limits of the NP Metaphor

Is the nature-preserve hypothesis a complete picture of the human/alien relationship? I doubt that it contains the whole answer. Yet like any metaphor of contact, it can help us to grow our understanding of the visitors. Their motivations and their paradigms of life are undoubtedly much different, just as ours are far removed from those of the animals. Thus, while not the whole story, the NP Metaphor might just offer us an important piece of the puzzle.

Occasionally the UFO phenomenon displays behavior that is very much at variance with what we would expect from "naturalists" conducting research. UFO intrusions into military airspace, the appearance of one or more UFOs over urban areas[ciii] and similar cases in which the phenomenon seems to take a higher profile, all would imply that this interaction is somehow more than simply scientific study. Or is it? Could these events simply be further examples of some form of scientific study we do not yet understand?

Combining the Models

How do the models we have discussed in this chapter mesh with experiencers' accounts of abduction and contact? Furthermore, how does the alien interest in human genetics, as described in chapter 5, fit in with such contact models? Might such genetic interest be part of some form of ET research, or "wildlife management" plan as previously suggested?

In the next chapter, we will combine these ideas to build a synthesis, an overall hypothesis of contact - the framework for the Cosmic Bridge.

Chapter 9 – Combining the Models

What are the visitors really doing here? In chapter 6, we discussed their interest in human genetics. In chapter 8, we examined some possible models of contact and the a possible extraterrestrial strategy for Earth. Is there such a strategy? If so, how can we discern what it is?

There are many theories about the nature and intentions of the visitors. However, we have noted many common themes in abduction accounts, suggesting that although diverse and paradoxical, the phenomenon has a deep self-coherence. In this chapter, we continue our journey toward an overall theory of contact, building The Cosmic Bridge.

Revisiting the Alien Agenda

How can we combine the ideas based upon what we have so far discerned about the phenomenon? Thus far, we have studied models[civ] of the phenomenon such as the Deardorff[cv] model, the Genetic/Hybrid model and the Zoo/Nature-Preserve model. We have further suggested that to the visitors it might somehow be vital to "civilize" humanity. In this chapter, we strive to unite these ideas into an overall hypothesis and examine such a theory in the context of the visitor experience. Finally, we look at ways in which we can test such a theory.

The Emergence of Humanity

To understand possible motivations and agendas for the visitor presence, let's take a step back and imagine that we can view humanity from their perspective.

1. Based upon the state of human civilization, as discussed in chapter 7, we can suggest that open contact could be very damaging to our culture. Thus, we concluded that humanity is not ready for contact.

2. Even a cursory look at history suggests that humanity is a fundamentally warlike being. In the present era, our largest single economic endeavor is war.[cvi] During modern times, our history has been dominated by such cataclysms as World War 1, World War 2 and uncountable genocidal massacres.[cvii]

3. Humanity has developed the capability for destruction on a planetary scale. We now have arsenals of multi-megaton nuclear weapons. Thus, we now have the capability of destroying our entire world - and by extension, other worlds as well.

4. Humanity may now be on the verge of developing interstellar flight.[cviii] Advances in zero point theory,[cix] the understanding of gravity and other areas of physics[cx] suggest that within a few hundred years we may have the capability to travel to the stars.

Given humanity's warlike nature, we can imagine those in the galactic civilization, which we presume to exist out there, asking whether they want humans venturing out into the cosmos. In humanity's current state of evolution, I can imagine that the answer is probably "No." Yet the recent rapid pace of human technological advance suggests the worrisome prospect of just such an expansion.

Many, including myself, believe that it is humanity's future to expand into space. Many in the space exploration community speak of a new manifest destiny: the destiny of humanity to expand to the planets - and eventually to the stars.[cxi] If humans are indeed destined to go to the cosmos, then by the time we realize such a future, will we have changed our culture and habits? Will we rise above our warlike ways? Or will we simply carry our wars with us to the stars?

This question suggests an immediate ET agenda: to prevent our warlike race from spilling out into the galaxy - to provide quarantine and containment until humanity is ready for contact. The neighbors might consider it vital that we somehow be changed, tamed or civilized, in order to avoid the universal catastrophe that unchecked human expansion might bring. We can therefore imagine James Deardorff's "ET Strategy for Earth" as a blueprint for human/ET relations, both now and far into the future. We can begin to understand that such an agenda would probably have a high priority for our cosmic neighbors.

Several experiencers, while deep in trance, have related to me their impression that indeed, this is an important goal of the visitors: to bring humanity to a higher level of civilization. Several have concluded that for this reason, the visitors seek to foster in humanity "a higher level of awareness."[24]

Might this correlate with the psychic awakening and perspective shift that I have observed in many abductees? Let's examine such awakening in the light of these ideas and see for ourselves.

---

[24] Note: An exact quote from one experiencer.

Human Awakening

An eastern parable asks, "How do you eat an elephant? - One bite at a time." Similarly, we might ask, how do you change the entire human race? - One person at a time.

Earlier, we discussed the phenomenon of psychic and spiritual awakening that often occurs to the individual experiencer. In this section, we ask whether this might be part of an overall drive to change humanity.

We noted that signs of the phenomenon often include a sudden shift in perspective to a more spiritual or metaphysical viewpoint. Accompanying this is the onset of psychic abilities and increased intuition. As more than one experiencer said to me, "It was as if my eyes were suddenly opened."

Earlier, we noted that the CE4 phenomenon is just one of many possible drivers of such emergence. However, the result is the same, a form of both personal growth and psychic connectedness. As I have interviewed them, many awakened experiencers have described such an increased sense of connection. For better or for worse, voluntarily or otherwise, they are expanding out with their minds, forming psychic bonds with others. Thus, as they grow they interlink, each one becoming a node in an awakening Network of a more interconnected humanity.

Barbara Marx Hubbard, in her book *Conscious Evolution*,[cxii] describes this psychic networking in terms of Stuart Kauffman's theory of networks and self-organizing systems[cxiii]. Hubbard describes how, as nodes of any network become progressively more interconnected, at some point a critical transition occurs. The fundamental nature of the network changes. It is a phase change, like the formation of a crystal. The random disordered behavior of individual elements becomes that of a coherent whole.

Hubbard theorizes that the ultimate result of progressive human psychic interconnection may be just such a transition - in which individual humans unite to become collectively awakened, a greater super-aware unity.

Many mental health and spiritual workers describe a psychic and spiritual awakening, which appears to be in progress.[cxiv] In addition, experiencers such as Carolyn, Amy and Jenny, whom we met in earlier chapters, describe these same experiences and the sense of interconnectedness with others. Might this be the early signs of such a phase change? Or to use the words of Barbara Marx Hubbard, could this be the start of the "conscious evolution?"

As described earlier by Dr. James Deardorff, perhaps the way to contact "primitive" beings such as humanity is from the bottom up, at the grass roots. Perhaps the way to ready us for contact is to foster the emergence of each person, for one person, two persons, then thousands, then millions - and eventually billions, building a critical mass toward the conscious evolution.

The "Indigo Hypothesis"

We have discussed the phenomenon of psychic and spiritual awakening that seems to occur for the individual experiencer. In addition, we observed earlier that the visitor experience seems to run in family lines. As I meet the families of experiencers, I have noticed that, in some cases, the characteristics emerging in newly awakened adults already seem to be present in their children. Many close encounter experiencers describe their children as gifted, psychic, deeply empathic and other characteristics similar to those of awakened adults. As an informal observation, I have noted a disproportionately large number of such children among these families. I have wondered about the mechanism by which these children are awakened. Is it due to influences of the visitors, or is there something else at work?

In chapter 6, we discussed the idea of genetic manipulation that seems to be a key focus of the abduction phenomenon. Budd Hopkins and David Jacobs argue that this is intended to create some form of hybrid being for purposes presently unknown to us. However, we can also ask whether such re-engineered genes might somehow be integrated into the overall human gene pool for purposes we do not yet understand.

One variant of this genetic hypothesis is the theory that such a hybridization process is intended to breed a new human. One possible representation of such an advanced human is described in the literature as the "Indigo Child." This term is based upon the work of Jan Tober and Lee Carroll in their book, "The Indigo Children."[cxv] Wendy H. Chapman also describes the Indigo Children on the website www.metagifted.org.[cxvi] Both works describe these children as being deeply empathic, visionary, intuitive and having a deep sense of mission and purpose. They seem to live with a thin veil between the physical and spiritual worlds. Furthermore, they[cxvii] are often intellectually gifted, resistant to authority and endowed with a high degree of psychic ability. They tend to have difficulties with the more mundane, non-creative aspects of life.

Is there a correlation between the visitor experience and the Indigo-child phenomenon? Might a "genetic agenda" of the visitor phenomenon result in an increased number of such children?

This is what I call the "Indigo Hypothesis": That the genetic changes fostered by the visitors might actually begin to appear in the general populace, in the new generation of humanity. We theorize hybridization changes by the visitors being assimilated into the overall human gene pool, resulting in the emerging generation of "Indigo Children" that we see today.

What the true purpose would be for such a program of "human re-engineering" we can only guess. Nevertheless, we can speculate that as these children take their functional place in society, we might somehow begin to see a more "awakened" conduct of human affairs. Perhaps we might see a new humanity, more ready for contact and more accepting of an ET presence on earth.

Integrating the Models

Ultimately, is there a primary agenda behind the visitor presence? We have so far considered a number of different models. Can we, with what we have learned to date, begin to synthesize an overall theory of the phenomenon?

In the previous chapter, we discussed possible ET strategies for contact with humanity. The most basic and all-encompassing model seems to me to be the Deardorff "Possible ET Strategy for Earth" - a hypothetical structured program of contact, which we can imagine that the visitors might follow. Such a model suggests a slow, steady program of observation, limited contact at the grass-roots level, with gradual and progressive revelation of themselves as humanity adapts to their presence. It also allows occasional limited intervention should a situation on Earth become serious enough to so warrant. However, the ultimate goal of such an agenda would be to prepare Earth for eventual open contact and disclosure. This would seem to be the most basic paradigm for much of the overall activity of the visitors. We can thus think of the Deardorff model as a framework for our combined hypothesis.

Within this framework, we can also see the nature-preserve (NP) metaphor as potentially being an accurate model of the on-going Human/Visitor relationship. We can envision a nature-preserve Earth in which extraterrestrial "naturalists" conduct research and perhaps manage the evolution of the indigenous human "wildlife." This model may serve as a framework to understand better what we now know as UFO abduction.

We have touched on the idea of humanity as a potential threat to the cosmic community. What can the visitors do to prevent a dangerous humanity from spilling out into the interstellar neighborhood? The obvious answer would seem to be to quarantine and somehow civilize humanity through contact and education. Ultimately, we can ask if it may actually be necessary to breed a more acceptable humanity. This suggests an immediate motivation for the genetic agenda as described by the Hopkins/Jacobs Model of UFO abduction - and leads us to an "Indigo Hypothesis."

At the individual level, the visitors seem to foster a psychic and spiritual awakening. Simultaneously, they seem to be breeding some form of apparently more desirable characteristics into a new race of human beings. This suggests that the process is intended to engender a new humanity more acceptable to the cosmic community, our preparation for contact.

Historical Considerations

As we indicated in chapter 7, various alternative theories of history, such as those of Zachariah Sitchin,[cxviii] suggest that humanity might have been a creation of extraterrestrial visitors.[cxix] Graham Hancock, in his book, *Fingerprints of the Gods,*[cxx] also argues that there is a hidden history of civilization, perhaps with heavy influence from extraterrestrial beings. While we must treat these ideas with discernment, we can imagine the scenario of a managed, guided Earth. Perhaps a cosmic super-civilization has nurtured our world throughout the ages. If this is true, then perhaps these beings have not yet completed their experiment and we are a work in progress.

Regardless of which (if any) alternative historical theories we choose to consider, we can see that ours is a planet in which "the Experiment called Humanity" seems to be reaching a critical phase. We can also suspect that, due to humanity's current rapid rate of technological advance, this phase of the experiment has taken on a new urgency.

Human advances such as nuclear power and weapons, space travel and communications suddenly threaten to bring humanity to the doors of the cosmos. Will humanity mature and step out of our home world into the cosmos, to be welcomed by our neighbors? Or will we become a threat to those very neighbors? Only time and we humans ourselves can determine the answer to these questions.

Human Potential

So far, in this chapter we have focused on the darker side of humanity. However, while we humans are often brash and barbaric, our history also shows a humanity that has a lot to offer. We are a driven and diverse species, a race of artists, thinkers, explorers and adventurers. Along with danger, we bring to the cosmos a thirst for the new and an insatiable curiosity - creativity, innovation and adventure.

In many UFO abduction accounts, experiencers describe how the visitors seem to lack many of these same dynamic characteristics. They describe how the alien realm seems to be sterile, featureless and stagnant. However, the greatest weakness of the visitors seems to be their inability to show emotion - love, passion, or any other of the heartfelt states of being that make us human.

In *Intruders*, Budd Hopkins is the first to describe this lack of emotion as a characteristic of the alien realm. He builds a case for at least part of the abduction phenomenon occurring to teach the visitors and their hybrid creations how to feel and display such emotions.

In the concluding chapter of *Intruders*, Hopkins suggests that this may be one reason behind the child presentation scenario in some abduction events, in which the experiencer is introduced to the hybrid child, which apparently is the result of earlier genetic/reproductive procedures done to them. The abductee describes feeling a powerful sense of love and the certain knowledge that this child is theirs. In addition, many experiencers such as Carolyn, described extensively earlier in this book, feel this powerful sense of love and longing for their "starchild" somewhere out there.

Do the visitors wish to bring such characteristics back into their own makeup? Is our ability to feel emotion; be it love, passion, or any other feelings, something that humanity can offer to the cosmos? If so then we can wonder if one day, humans might be the ones to bring a new spark of life to a stagnant cosmic community. Is this a spark that perhaps only a brash young humanity can provide? We can ask if, perhaps for this reason alone, humanity might be a valuable asset to the cosmic community, one very much worth saving. Perhaps it is in the interests of the cosmic community to see humanity survive and to one day take its place among the stars.

Is the UFO abduction phenomenon an ET effort to somehow foster our survival, perhaps to guide (or breed) us into the future? We can imagine that it would be a delicate task: to civilize humanity to the point where we can move beyond our warlike ways, while at the same time preserving the passion and drive that make us human.

Perhaps we can picture this massive change as a vast covert "Human Improvement Project." And we must realize that the results of this effort are still pending.

Testing the Hypothesis

The purpose of a hypothesis is to serve as a model of the world from which we can derive testable predictions. So what would be a good test of our combined hypothesis? How can we determine whether the visitors are in some way altering the foundations of our society, even our very being? Though there are probably many ways, here we consider some tests to establish the validity or non-validity of our combined model.

1. Genetics: Might the genome of Indigo children[cxxi] show common markers not present in non-indigos? If so, then what characteristics would these genes encode? If we didn't detect common but unique genetic markers among indigo children, would this support some form of null hypothesis?[cxxii]

2. Genealogy: Does a comparison of the family lines of Indigo children with those of close encounter experiencers show a correlation? Are Indigo children more often found in the family lines of experiencers? If so, what does this imply? Would a direct correlation indicate that the UFO/CE4 phenomenon and the Indigo children were causally linked?

3. Psychic Abilities: Do Indigo children have greater psychic abilities? Perhaps this can be tested with experiments such as those at the PEAR lab at Princeton University,[cxxiii] which has been developing scientific tests of ESP and non-local consciousness.

4. Conscious Evolution: Do we see an overall increase in the psychic abilities of humanity, perhaps as measured by various field consciousness experiments?[cxxiv] This might be an indication that some form of aggregate human mind-field was present and strengthening, perhaps an indication of an overall

psychic awakening of humanity as described in Barbara Marx Hubbard's book *Conscious Evolution.*

5. Effects on World Affairs: We can carefully observe cultural, scientific and political events over the next twenty-plus years as the hypothesized new generation of Indigo-Children becomes active in society. As this generation reaches a political-social critical mass, what effect does this have on the conduct of our affairs? Is there an incorporation of intuition and spirituality into mainstream society and politics?[cxxv] Or might there be other unforeseen effects, positive or otherwise?

6. Open Contact: Assuming that the above changes are taking place, do we then observe an increase in multi-witness UFO sightings and increasingly open contact events over the next few years?

In the next decade or two, will we begin to see a fundamental change in humanity? If so, will we find contact and UFO sightings becoming more commonplace? Will the visitor experience broaden and become more overt? In current events, do we see any suggestions of the tests predicted by our hypothesis? The answers to these questions are still to come, but they will provide a crucial test of our ideas on the agenda behind the visitors' presence.

Toward a Combined Hypothesis

What are the visitors doing here? We began this chapter by asking this question. In earlier chapters, we examined different theories of contact and in this chapter, combined them to build an overall theory of the visitor presence. We combined the idea of genetic manipulation from chapter 6 with the idea of psychic and spiritual emergence. We then developed the idea that a grassroots awakening of humanity is beginning. We also examined possible reasons the visitors may have for coming here and concluded that it may be to prevent a warlike humanity from spilling out into the cosmos at large.

If our combined hypothesis is true, then to me this suggests that the visitors want us to grow and eventually join the cosmic community. And in this idea, I find a great sense of optimism. To me it offers hope that humanity can indeed become an evolved star-faring civilization. Perhaps there is a cheering section out there rooting for us. Perhaps we have celestial neighbors awaiting our arrival over the Cosmic Bridge.

Chapter 10 - Crossing the Cosmic Bridge

Throughout this book, we have built an expanding picture of our relationship with the visitors. We have asked what their strategy might be for contact with worlds like ours. How does this show up in the experiences of abductees and contactees?

In the first half of this book, we examined UFO encounters and the experiencer's individual relationship with the visitors. In later chapters, we broadened this view to consider the visitors' overall strategy for contact with Earth. In the last chapter, we looked at the reasons why the visitors may be paying so much attention to our world. We suggested that there might be a "human improvement project" underway, to groom us for future contact, part of an effort to bring humanity into the cosmic fold. We suggested that due to the current rapid advance of humanity, this project might have a high degree of urgency to it.

One thing that bears review is the situation in the human realm. We earlier established key danger points regarding the current state of humanity. A brief look at contemporary events immediately suggests why the visitors may be paying so much attention to us in the present day:

1. As we discussed in chapter 7, the human psyche is probably not yet prepared for overt extraterrestrial contact. We noted the possibility of significant damage to human culture and suggested that any enlightened E.T. civilization would choose to avoid this.

2. We noted that in the present day, humanity's largest single endeavor is war. This behavior would make us, at best, unpleasant neighbors to the cosmos.

3. For the first time in the history of our civilization, humanity has developed the capability for destruction on a planetary

scale. Thus, we noted that in the near future it may be possible for us to pose a direct danger to our cosmic neighbors.

4. Space exploration is becoming a major endeavor of the economically developed nations. We are planning longer-term expeditions to the Moon and Mars, as well as robotic ventures to the outer planets. In addition, a quick look at the leading edge of physics suggests that we may be closer than we think to the advent of star travel. Perhaps within a couple of centuries, such deep space journeys might become reality.

Is our civilization developing into a potential threat to the cosmic community? Will we carry our warlike ways into the heavens? Perhaps the scenario of a warlike humanity spilling out into the cosmos is a real possibility.

Alternatively, can we learn to contain our darker impulses, to live with one another and with the Earth in a more harmonious, constructive relationship? I often wonder if those of the cosmic community, whoever they may be, feel a responsibility help us to follow the latter path.

Could this be the prime reason for the visitors' interest in our world? Are the visitors trying to intervene in a covert way, to alter our nature before we humans begin traveling to the stars? I suggest that, to any cosmic-scale civilization, the state of humanity MUST be a matter for concern.

In nearly the same thought, we again note the profound shift of awareness occurring among experiencers. We have seen how spiritual and psychic awakening is often a powerful component of their lives. We have also seen how the UFO phenomenon and the awakening phenomenon seem to reinforce each other. In chapter 4, we noted that for many experiencers, it seems to force them into a fundamental shift of worldview.

We also suggested in the last chapter, that psychic and spiritual emergence might be reaching a critical mass, beginning what Barbara Marx Hubbard termed a conscious evolution. We can speculate that perhaps such a fundamental shift within humanity is a key purpose behind the visitors' presence.

We also noted in the last chapter that many children of experiencer families seem to be Indigo children - brilliant, precocious, psychic, and philosophically advanced. We have suggested that they may represent a new humanity, more compatible with the cosmic community. Therefore, we asked whether the visitors might actually be fostering these changes. Could this be part of their program to prepare humanity for contact?

We noted James Deardorff's theory that Earth is under quarantine until we have evolved enough for acceptance into the cosmos. Are the changes we have discussed here the first signs of such an evolution?

Slowly but surely, as the awakening takes hold, I can imagine us closing the gap between our cosmic neighbors and ourselves. I propose that it is human destiny to be in such a deepening relationship with the visitors.

Perhaps the primary reason for the UFO abduction phenomenon is to alter humanity - to force our psychic maturation before we begin to move out among the stars. If this is true, then as these efforts progress, we may see ever more psychic and spiritual awakening. We may even see a fundamental change in the way we humans conduct our affairs. As we do, will we also see an increasingly overt alien presence? Perhaps as they near their objective, we can look forward to the opening of both human and cosmic civilizations to contact.

A bridge seems to be in the process of formation - the convergence of humanity with the reality of the visitors. Given the pace of human events and what they may represent to the visitors, I suggest that, both for them and for us, there is urgency to this endeavor.

At the level of the individual and of humanity as a whole, I can imagine a future in which our relationship with the visitors becomes ever closer. We noted how experiencers are a bridge between worlds. They often live with one foot in each realm - the human and the phenomenal. As the awakening grows in scope, approaching critical mass, I suggest that this bridge will become universal.

At that point, we can picture contact becoming large-scale and overt, humanity increasingly becoming part of the cosmic community. We can imagine it as the ultimate unity, humanity and the cosmos coming together on the Cosmic Bridge.

Appendix A – Research and Support Organizations

MUFON

The Mutual UFO Network (MUFON) is a grass-roots organization of dedicated researchers, investigators, and interested parties attempting to understand the phenomenon of UFOs and close encounters. MUFON specializes in the collection and analysis of worldwide sighting and encounter reports.

In the U.S., it is estimated that up to 10 percent of the population have seen something in the sky at least once in their lives, which they could not explain. It is also estimated that at most 10 percent of all such sightings are reported. While reports of close encounters are frequently handled differently from sighting reports, all contribute in important ways to the overall understanding of the UFO and close encounter mystery.

The focus of MUFON is to perform as detailed an investigation as possible of reports of UFO sightings and encounters. Sightings may be reported through direct channels such as the MUFON on-line sighting report form. At other times a report may simply be passed along informally by word of mouth to an investigator. In all sighting cases, it is the goal of the Mutual UFO Network (MUFON) to increase the percentage of such sightings that are reported and analyzed. It is our hope that by analyzing such reports, we may gain a clearer picture of the UFO phenomenon.

MUFON maintains a site on the internet World Wide Web, with an on-line sighting report form, at its website: www.mufon.com. When a sighting report is received, based upon the location of the reporting witness, the MUFON staff will then notify the state director or one of the field investigators in the region who will begin a detailed investigation.

In addition, the Minnesota chapter of MUFON maintains a website where regional sightings may be reported (www.mnmufon.org). From that site, the witness can report a sighting, or can look up the name and phone number of (or send e-mail to) one of our local field investigators.

The initial purpose of a UFO field investigation is to determine if the object the witness observed was in fact an unidentified object. Once a sighting is reported, MUFON investigates sighting reports in detail. The task of investigation is assigned to one of the field investigators in MUFON. That investigator then contacts the witness and, if the witness feels comfortable further describing the sighting or encounter, arranges an interview.

During a sighting witness interview, the investigator will ask the witness to describe the UFO event from beginning to end. The investigator may also ask a number of questions in order to bring out as much information as possible about the sighting. The investigator will fill out a sighting report, which will be sent to MUFON headquarters. Also, either before or after the interview, the investigator will gather as much additional corroborating information as possible. Whenever possible, this will include items such as weather data, a detailed survey of the sighting area and any information from law enforcement regarding other possible reports of unexplained phenomena at the time of the event.

If a sighting does not seem to have a conventional explanation, the investigator will send a detailed report to MUFON headquarters to be included in the MUFON sightings database. In addition, further research may be done by the investigator, or by one of the MUFON consulting experts.

The field investigator's report, along with the sighting report from the witness, will be logged anonymously into the MUFON UFO Database. The information, with the witness's name and other identifying details removed, will be available to UFO researchers for future scientific study.

From the investigation of many sightings, researchers can discern overall characteristics of UFOs, trends in sightings and other aspects of the phenomenon.

Each UFO sighting is a piece of the puzzle and any given sighting may contain an important clue. I urge any person who has seen a UFO, or has had a UFO-related experience, to report it. Reports may be sent to one of our investigators, through the Minnesota MUFON web site, or to the report form on the MUFON website. From this, we can work with you to perform a detailed followup investigation.

MUFON can be reached at the following location:
Mutual UFO Network, Inc.
P. O. Box 279
Bellvue, CO 80512-0279
(970) 221-1836
e-mail: **HQ@mufon.com**
http://www.mufon.com

We are actively seeking reports of UFO experiences. Your experience could be a key to helping us understand this mystery. If you wish, we will keep all information confidential. Please call us; your sighting/encounter is very important.

Close Encounter Research Organizations

The following is a list of additional organizations that research close encounter reports, and/or can assist the experiencer in coming to terms with the close encounter phenomenon:

- PEER (Program for Extraordinary Experience Research)
  John E. Mack Institute
  PO Box 398080, Cambridge, MA 02139
  617-497-2667
  info@johnemackinstitute.org
  http://www.johnemackinstitute.org/

- Intruders Foundation
  P.O. Box 30233 New York, NY 10011
  212-645-5278
  e-mail: ifcentral@aol.com
  http://www.intrudersfoundation.org

- ICAR (International Center for Abduction Research)
  12 West Willow Grove Avenue 191
  Philadelphia, PA 19118
  http://www.ufoabduction.com

The names and contact information for other, additional organizations, researchers and/or therapists may be available by contacting MUFON.

Appendix B - Types and Kinds of UFO Events: The Vallee Classification System

What is a UFO? The term tickles the imagination. We wonder what they are and then realize that they are unknown by their very definition. A UFO - Unidentified Flying Object - is an object in (or from) the sky which at the time of the sighting could not be identified. A UFO event can range in strangeness from the simple sighting of a distant object to the ultimate experience of contact with the extraordinary. Each event offers us one point within a vast spectrum of mystery.

Such a tremendous variety of sightings and encounters presents a staggering task to the researcher. Where do we begin? How can we ask questions about them? How are we to describe them? What are we even to call these events?

One of the first stages in the scientific investigation of any phenomenon is to somehow characterize and classify it[cxxvi]. The UFO mystery is no exception. What are the patterns of the phenomenon? What do the many UFO sightings have in common and how are they different? What models or theories can we develop to help us understand this phenomenon?

In the 1970s, Dr. J. Allen Hynek[cxxvii] developed a system of classification of UFO sightings based upon key characteristics of the event. In this system, sightings and encounters were classified as:

- Nocturnal Lights (NL)
- Daylight Disks (DD)
- Radar/Visual (RV)
- Close Encounters of the First Kind (CE1): sighting of an object in relative proximity to the witness
- Close Encounters of the second kind (CE2): Physical evidence left behind by a UFO event.
- Close Encounters of the third kind (CE3): the sighting of, or interaction with, entities associated with a UFO.

Subsequently, J. Allen Hynek's system was extended by Dr. Jacques Vallee[cxxviii] to classify UFO and paranormal events in general. Just as in the Hynek system, the Vallee system organizes events according to the type or essential nature and the kind, or degree of meaningfulness, of the event. However, Vallee generalized the categories to include overall consideration of the event type and the degree of strangeness of the event.

The Type denotes the essential character of the experience for the witness; ranging from the simplest of distant sightings to the strangest anomalies and close encounters.

Types of events as described by Jacques Vallee are:
- The Distant Encounter - Flyby (FB) or Maneuver(MA): This is the sighting of a UFO in the distance. The object may simply fly past the witness, or it may undergo some maneuver that suggests artificial control. Sightings of this type are more likely have non-extraordinary explanations such as a misidentified aircraft, an astronomical body (e.g. Venus, Sirius, etc.) or some other prosaic phenomenon. Thus, in most cases, the UFO becomes an IFO (Identified Flying Object). By far the majority of all UFO sightings are such distant, often-explainable events.
- The Close Encounter (CE) - A sighting of a UFO or related phenomenon in close proximity, in which detail is visible: Sightings occurring at closer range or under circumstances of higher strangeness are less likely to have conventional explanations. It is a rule of thumb that the closer the range at which a UFO is seen, the less likely it is to be explainable in prosaic terms.
- The Anomaly (AN): an anomaly is a (probably) non-UFOlogical event that defies a conventional explanation. Examples of such events may be the sightings of ghosts or monsters, the occurrence of religious miracles, out of body experiences, etc. Cases in this book and in the UFO/CE4 literature show that close encounter experiencers frequently tend to experience such anomaly events. Is there a connection

between anomalies and close encounters? Might the understanding of anomalies be a key to understanding the human-Visitor interaction?

As mentioned above, for any sighting or experience there will be associated, some degree of strangeness or meaningfulness. How close was the object? How much interaction was there between the witness and the phenomenon? What was the effect on the witness? The degree and character of this relationship is represented in the "kind" of event.

Kinds of Events as described by Jacques Vallee are:
- First Kind - An object or phenomenon is seen apart from the witness where no overt interaction occurs: The witness remains separate from the phenomenon (though as we see in chapter 1, there may be a subtle relationship between the two).
- Second Kind - Physical evidence is left behind by the phenomenon: Although I mention it here for the sake of completeness, I leave the fascinating study of physical UFO evidence to other works.[cxxix]
- Third Kind - Entities are observed by the witness: The witness remains separate from the phenomenon but observes some type of entity. Although the witness is an observer, we again note that interaction with the phenomenon may be increased.
- Fourth Kind - There is interaction and/or participation by experiencer: There occurs a transformation of the experiencer's reality into the reality of the phenomenon (most typically, a UFO contact or abduction).

The simplest and by far the most common type of sighting is that of the distant encounter of the first kind, an event in which a witness makes a simple observation of something in the sky. For example, a distant object flying past, which may be unidentified at that moment, would be referred to as a flyby of the first kind, or FB1.

If the UFO appears to maneuver, suggesting intelligent control, but remains distant from the observer, then the sighting becomes a Maneuver of the First Kind or MA1.

When an object approaches to the point in which detail is visible, then the event becomes a close encounter, a CE1 or Close Encounter of the First Kind. Where an entity is visible, in connection with a UFO, then the event becomes a close encounter of the third kind, or CE3. If it is apparent that an entity is visible, but no UFO is observed - such as during a ghost or monster sighting - then an anomaly has occurred - an AN3, or anomaly of the third kind.

In each case, the level of "strangeness" increases as the "kind" of the event increases. The relationship between the witness/experiencer and the object becomes closer and a deeper flow of meaning occurs. We find that the bidirectional interaction of the relationship also increases and the effect on the witness becomes progressively more profound.

The ultimate case is that in which the witness becomes a (usually unwilling) participant - this event is now an event of the fourth kind. At this stage the sighting has now become fully experiential, an abduction, contact or other change in reality. The witness' reality is transformed into that of the phenomenon.

If this event involves a UFO, then the event becomes a close encounter of the fourth kind, or CE4. The witness has become an experiencer - willing or otherwise. He/she has become a voyager across The Cosmic Bridge.

About the Author

Craig R. Lang is a certified hypnotherapist with the National Guild of Hypnotists. He lives and has his hypnotherapy practice in Brooklyn Center (Minneapolis), MN. In addition to hypnotherapy, he is a field investigator with the Mutual UFO Network and researches UFO sightings, anomaly reports and close encounters.

His full-time job is in software engineering with a medical electronics company in St. Paul, Minnesota. In addition, his other interests are astronomy, parapsychology, creative writing, spirituality and meditation, plus various outdoor sports. Craig can be reached via his website at: www.craigrlang.com, via e-mail at craig@craigrlang.com, or by phone at 763-257-7334.

Notes and References

Introduction
[i] MUFON, abbreviation for:

Mutual UFO Network, Inc.
P. O. Box 279
Bellvue, CO 80512-0279
(970) 221-1836
e-mail: **HQ@mufon.com**
http://www.mufon.com
(See also Appendix A, Close Encounter Support and Research Organizations)

Chapter 1
[ii] http://www.nidsci.org/news/roper_surveys.html
Discussion of the 1991 and 1998 Roper Poll of Unusual Personal
Experiences
Notes: The criterion for being an experiencer is experiencing 4 of 5
indicator phenomena more than once. The overall estimate of the 1998
poll is that approximately 1% of the population fit this criterion. The
number experiencing all indicator phenomena appears to be
approximately 0.2%.
5% to 10% of the population indicates that they have had at least one
UFO sighting in their life.

[iii] The description of the Medicine Lake Triangle sighting can be found on the
Minnesota MUFON website at: http://mnmufon.org/99-1i.htmlto

Chapter 2
[iv] David Jacobs,
- Secret Life, Touchstone, 1993
- The Threat, Simon & Schuster, 1999

[v] Raymond Fowler, The Watchers, Bantam, 1991

[vi] John Mack, Passport to the Cosmos, Three Rivers Press, 2000

[vii] Alien Discussions, Proceedings of the Abduction Study Conference at MIT
Editors: David E. Pritchard, John E. Mack, 1994

[viii] Raymond Fowler, <u>The Watchers</u>, Appendix B, describes in detail, the prototypical abduction scenario.

[ix] Note: In close-up UFO sightings (an apparent CE1 event), witnesses may feel as if they are the apparent focus of the phenomenon. When such an intentional focus occurs, especially if there is a logical discontinuity or paradox in the sequence of events, investigators consider it likely that an abduction occurred.

[x] David Jacobs, <u>The Threat</u> and Budd Hopkins, <u>Intruders</u> and <u>Missing Time</u> both contain extensive discussion of screen memories and other related side effects of the close encounter phenomenon.
See also Bryant and Seebach, <u>Healing Shattered Reality</u>

[xi] Bryant and Seebach, <u>Healing Shattered Reality, Understanding Contactee Trauma</u>, Wildflower Press, 1991

[xii] Anomalous pregnancies are described extensively in collected works of Fowler, Hopkins and Jacobs. They are also extensively described by abductees such as Carolyn and Jenny, referred to in this chapter.

Chapter 3

[xiii] Budd Hopkins, <u>Intruders</u> documents physical evidence associated with a UFO sighting, and abduction. In addition, Hopkins' book "Witnessed" documents multiple witnesses to a UFO sighting associated with an abduction.

[xiv] Raymond E. Fowler, <u>The Andreasson Affair</u>, Prentice Hall, 1979.
This account of Betty Andreasson Luca's close encounters clearly and superbly documents multiple witnesses, physical evidence and the mental/metaphysical aspects of the CE4 experience

[xv] Jenny Randles, <u>UFO Reality</u>, R. Hale, 1983

[xvi] David Jacobs, "Thinking Clearly about UFO Abduction," on www.ufoabduction.com:
        http://www.ufoabduction.com/ICARSite/thinking1.htm

[xvii] <u>Flying Saucers : A Modern Myth of Things Seen in the Skies</u> by C. G. Jung, Republished by Bollingen, 1979

[xviii] Jacques Vallee, <u>Passport to Magonia</u>, McGraw Hill, 1993

[xix] Budd Hopkins and Carol Rainey, Sight Unseen, Atria, 2003
Note: this describes instances of invisibility associated with the UFO/CE4
phenomenon

[xx] Donna Higbee describes this phenomenon on web site:
http://members.aol.com/rapunz1/invisibility.html

[xxi] David Jacobs, "Thinking Clearly about UFO Abduction," on
www.ufoabduction.com:
        http://www.ufoabduction.com/ICARSite/thinking1.htm

[xxii] Unusual Personal Experiences: 1991 and 1998 - National Institute of
Discovery Sciences (http://www.nidsci.org/news/roper_surveys.html)
Discussion of the 1991 and 1998 Roper Poll of Unusual Personal Experiences
Notes: The criterion for being an experiencer is experiencing 4 of 5 indicators of
the phenomena more than once. The overall estimate of the 1998 poll is that
approximately 1% of the population fit this criterion.

Note that this number is actually quite ambiguous. Depending upon how one
selects within the Roper poll results, the criteria for being an experiencer, the
number of experiencers can range from 0.1% to 1.0% of the population.

In addition, it is not clear from the Roper poll data whether or not the abduction
phenomenon is expanding in scope (the poll suggests that it may even be
decreasing as a percentage rate of humanity). However, David Jacobs and others
claim that this the phenomenon is expanding. Thus, trends in the scope of the
phenomenon appear to be a key ambiguity in our understanding of UFO
abduction and contact.

Chapter 4

[xxiii] "Unusual Personal Experiences: 1991 and 1998" - National Institute of
Discovery Sciences on the web at: www.nidsci.org/news/roper_surveys.html
Discussion of the 1991 and 1998 Roper Poll of Unusual Personal Experiences
Notes: The criterion for being an experiencer is experiencing 4 of 5 indicators of
the phenomena more than once. The overall estimate of the 1998 poll is that
approximately 1% of the population fit this criterion.

[xxiv] Bryant and Seebach, Healing Shattered Reality, Understanding Contactee
Trauma, Wildflower Press, 1991

[xxv] John Mack, Passport to the Cosmos, Three Rivers Press, 2000

Chapter 5

[xxvi] Whitley Strieber, The Secret School
This book describes a series of lessons, which Whitley Strieber received as a
child, at the hands of the visitors.

[xxvii] Barbara Harris Whitfield, Spiritual Awakenings, Health Communications,
1995

[xxviii] Peter A Sanders ,Jr.., You are Psychic, Fawcette Columbine, 1989
Note: I have found that this book is one of the better books on psychic
development. It contains a number of well-defined exercises to help develop
parapsychological abilities.

[xxix] Books by David Jacobs: Secret Life and The Threat , as well as his article:
"Thinking Clearly About UFO Abduction," presented at the MUFON 1998
International Symposium and published on Dr. Jacobs' website:
www.ufoabduction.com.

[xxx] A number of researchers have suggested that undetected CE4 experiences
may have occurred but might still be blocked from memory. I cannot deny that
this could be the case. However we explored as deeply as was ethically
acceptable to do, with no indication of CE4 activity. To explore deeper might
have generated the implicit hypnotic suggestion that the experiencer should "go
find an abduction event." In the interests of good therapy and research, we
stopped before this point. Based upon the lack of any indication to the contrary,
is my belief that in many/most of these cases, there was no CE4 activity to be
found.

[xxxi] Kenneth Ring, The Omega Project, Quill, 1993

[xxxii] Note to the reader: I acknowledge that contactee claims are greeted with
extreme skepticism by many within the UFO research community. However, in
this book, we will assume that the events indicated by the contactee, even where
controversial, were experienced as described.

[xxxiii] Lyssa Royal and Keith Priest, Preparing for Contact, Light Technology
Publishing, 1994
An example of contact literature which details information ostensibly received
from extraterrestrial entities through encounters and via channeling sessions.

[xxxiv] Barbara Marx Hubbard, Conscious Evolution, New World Library, 1998
An example of literature describing increasing frequency of spiritual awakening
Note: Another book is "Spiritual Awakening" by Barbara Harris Whitfield ,
described earlier

[xxxv] David Jacobs, The Threat describes the increase of the close encounter
phenomenon in the world, and projects that this is due to a cascade down the
entire family line of experiencer families. I have noted claims of this trend, but
found it to be less clear than Jacobs describes. Trends in close encounter
frequency is a current area of research/investigation.

[xxxvi] Lisette Larkins, Calling on Extraterrestrials, Hampton Roads, 2003
This is one of three books by Lisette Larkins, the other two being "Talking to
Extraterrestrials" and "Listening to Extraterrestrials"

[xxxvii] Richard Haines, CE5 – Close Encounters of the Fifth Kind, Sourcebooks,
1999

[xxxviii] CSETI: the Center for the Study of ExtraTerrestrial
Intelligence.(www.cseti.org)

[xxxix] Steven Greer, "Extraterrestrial Contact: The Evidence and Implications,"
Crossing Point, 1999

[xl] "The Indigo Hypothesis," by Craig R. Lang, published in the February 2004
MUFON UFO Journal

[xli] PMA Atwater, The New Children and Near Death Experiences, Bear & Co.
2003

[xlii] The relationship between UFO abduction and the Near Death experience is
extensively explored in the book, The Omega Project, by Kenneth Ring

[xliii] Raymond Fowler, The Andreasson Affair, Bantam, 1979

[xliv] Note: This is powerfully described in the book Healing Shattered Reality by
Bryant and Seebach

[xlv] David Jacobs. Secret Life, (see earlier references)

[xlvi] Raymond Fowler, The Synchrofile, iUniverse, Inc., 2004

[xlvii] Budd Hopkins address to the 1997 MUFON UFO Symposium, in Proceedings of the MUFON UFO Symposium, 1997

[xlviii] Bryant and Seebach, Healing Shattered Reality, Understanding Contactee Trauma, Wildflower Press, 1991

Chapter 6

[xlix] Budd Hopkins, Missing Time, and Intruders both build an excellent case for the genetic/reproductive aspects of the abduction phenomenon

[l] Budd Hopkins, Sight Unseen, and David Jacobs, The Threat, both extensively describe the alien interest in human genetics.

[li] Budd Hopkins book, Intruders and Raymond Fowler's book, The Watchers both extensively describe both the topic of hybrids and of child presentations.

[lii] Raymond Fowler, The Watchers, contains extensive descriptions of these incubation tubes containing apparent fetal hybrid beings

[liii] David Jacobs, The Threat builds a case for a non-benevolent purpose of the abduction phenomenon

[liv] John Mack, Abduction, Human Encounters with Aliens, Ballantine, 1995
Bryant and Seebach, Healing Shattered Reality (see earlier reference)

[lv] In Chapters 12 and 14 of Sight Unseen, Carol Rainey provides a list of transgenic experiments.

[lvi] Examples of institutions conducting transgenic research are:
- The Stanford School of Medicine, Transgenic Research Center, described at http://med.stanford.edu/transgenic/
- International Crop Research Institute for Semi-Arid Tropics (ICRISAT) conducts transgenic research on food crops. See http://www.icrisat.org

[lvii] Hopkins and Rainey, Sight Unseen (see previous reference)

Chapter 7

[lviii] Some excellent books with various views on potential contact scenarios:
- Timothy Goode's book: Alien Update, Avon Books, 1995
- Carl Sagan: The Cosmic Connection, Cambridge University Press, 2nd Edition, 2000
- F. Joe Lewels: The God Hypothesis, Granite Publishing, 1997
- John Mack: Passport to the Cosmos (as described above)

[lix] Two organizations which are actively involved in SETI are the SETI Institute (http://www.seti.org) and the SETI League (http://www.setileague.org)

[lx] See Interstellar Travel, Looking to the future at the website location: http://www.ac.wwu.edu/~vawter/StudentSites2002/interstellar_travel/current.html
This site discusses some of the difficulties and possibilities for interstellar travel. Another excellent book on the problems and possibilities of star travel is The Starflight Handbook by Matloff and Mallove, Wiley, 1989

[lxi] Seth Shostak, Sharing the Universe, Berkeley Hills, 1998
Frank Drake and Dava Sobel, Is Anyone Out There?, Delta, 1994
Car Sagan and I.S. Shklovskii, Intelligent Life in the Universe, Emerson Adams Press, reprint 1998

[lxii] Note: In the SETI paradigm, the term "contacted" means that detection is made of radio signals emitted from the home world of the civilization. Thus, contact means radio contact; technology means radio-technology, etc.

[lxiii] Making Contact Edited by Bill Fawcette, William Morrow & Company, Inc. New York, 1997
Section 3 of this book provides a summary of the factors involved in the drake equation.

[lxiv] We can think of the entire SETI endeavor as an experimental test of the Drake equation and the assumptions that accompany estimates of the number of radio-capable civilizations.

[lxv] An excellent discussion of the Fermi Paradox can be found on the UFO Evidence website at the location:
http://www.ufoevidence.org/documents/doc287.htm

[lxvi] Carl Sagan, Pale Blue Dot, Ballantine, 1997

[lxvii] An excellent summary of the current state of the search for extrasolar planets can be found at: http://exoplanets.org/index.html

[lxviii] For a good overview of the exploration and search for life on Mars, check out the NASA/JPL Mars website at: http://mars.jpl.nasa.gov/

[lxix] "Geologic Evidence for the Antiquity of Life" an MIT paper on-line (PDF format) at:
http://ocw.mit.edu/NR/rdonlyres/Earth--Atmospheric--and-Planetary-Sciences/12-007Spring2003/6677344D-7A6A-457D-A38B-95FACCCFCE67/0/Lec03.pdf

[lxx] An interesting paper, published on the Cambridge-conference network, discusses the lifespan of civilizations, using several earth civilizations as examples both of very short-lived and very long-lived cultures. This paper can be found on-line at:
http://abob.libs.uga.edu/bobk/ccc/ce110999.html

[lxxi] Michael Lindemann is futurist and journalist.
He writes, "A journalist deals in current news, as or just after it happens -- unlike a historian, who has the luxury of time to get fine details verified. A futurist studies events and trends that may shape the future, attempting to discern the nature of possible futures and, sometimes, advising on alternative paths that might be taken toward preferable futures."

He is the author of many articles on UFOs and extraterrestrial contact, including "Faces of the visitors" which can be found on www.cfree.org. He is a writer and the host of the movie UFOs and the Alien Presence, and of the book UFOs and the Alien Presence; Six Viewpoints (2020 Group visitors Investigation Project, 1991).

Information about Michael Lindemann and a summary of the views, including those which he presented to Minnesota MUFON can be found on the researchers pages of the UFO Evidence website at:
http://www.ufoevidence.org/researchers/detail100.htm

[lxxii] Much of this model is based upon ideas from the 10/9/99 talk at MN MUFON by Michael Lindemann regarding contact scenarios, entitled "A Future Full of ETs: Real Scenarios of Contact." Much of the material in this section is derived from Mr. Lindemann's talk, and from additional private e-mail correspondence. Where indicated, the material is directly quoted with his permission.

[lxxiii] NASA Origins Program: http://origins.jpl.nasa.gov/index1.html

[lxxiv] Mars Exploration: Search for Life page on the nasa.gov website: http://mars.jpl.nasa.gov/science/life/

[lxxv] Universe Today, April 2004: "Is there life on Europa?"
Online at:
http://www.universetoday.com/am/publish/is_life_europa.html?542004

[lxxvi] Allen Hills "Mars Meteorite" ALH84001, description in "Wikipedia":
http://en.wikipedia.org/wiki/ALH84001

[lxxvii] Possibility of life on Europa:
- "New Evidence for an Alien Ocean"
  http://science.nasa.gov/headlines/y2000/ast28aug%5F1.htm
- Richard C. Hoagland's original, unedited, highly controversial 1980 "Europa Proposal" in Star & Sky Magazine, Jan. 1980, reproduced on http://www.enterprisemission.com/europa.html

[lxxviii] Carl Sagan, The Cosmic Connection, Cambridge University Press, 2nd Edition, 2000

[lxxix] Robert Temple, The Sirius Mystery, Destiny Books, 1998
This book examines possible contact between the extraterrestrials and the Dogon and other peoples at various times in antiquity.
An excellent listing of works on the Dogon/Sirius Mystery can be found at:
http://www.ufoevidence.org/topics/Dogon.htm

[lxxx] John Carpenter, MSW, paper entitled "Centuries of Contact" (unpublished)
This paper, written by UFO researcher John Carpenter, summarizes evidence regarding possible extraterrestrial influence and contact indications in the prehistoric and historical record.

[lxxxi] Rev. Dr. Barry Downing, The Bible and Flying Saucers, Marlowe & Co, 1998

[lxxxii] Raymond Fowler, The Watchers, Bantam, 1991

[lxxxiii] Zachariah Sitchin is the author of multiple books which theorize that humanity has had extensive interactions with extraterrestrials which have subsequently been taken to be gods. His books include:

- The Twelfth Planet
- The Wars of Gods and Men
- The Lost Realms
- The Stairway to Heaven
- Genesis Revisited
- The Earth Chronicles Expeditions

Overall information about the work of Zachariah Sitchin can be found on http://www.sitchin.com/

[lxxxiv] The Brookings Report on Peaceful Uses of Space contains the section "The Implications of a discovery of extraterrestrial life." A copy of this report is on the Enterprise mission website at:
http://www.enterprisemission.com/brooking.html..
At the time of this writing, The Brookings Report is still seen as an authoritative indication of the potentially negative consequences of overt contact.

[lxxxv] "The UFO Phenomenon and the Suicide Cults – An Ideological Study," Budd Hopkins, published in the MUFON UFO Symposium Proceedings, 1997

[lxxxvi] The SETI Declaration of Principles is located on the SETI Institute website at: http://www.seti.org/site/pp.asp?c=ktJ2J9MMIsE&b=179287

[lxxxvii] The CSETI CE5 Initiative is described on the CSETI website: at the location: http://www.cseti.org/ce5.htm

[lxxxviii] "When UFOs Arrive:, Popular Mechanics, February, 2004
http://www.popularmechanics.com/science/space/1283081.html

[lxxxix] The possible crash of a UFO in Kecksburg, PA is described on the web at http://ufocasebook.com/Kecksburg.html

Chapter 8
[xc] The website UFOEvidence.org contains many articles on interactions between the United States military and UFOs. One such report is on UFO incursions over nuclear facilities is located at: http://www.ufoevidence.org/topics/nuclear.htm. Another report, on the UFO incursion over Malmstron Air Force base in 1967, is located at: http://www.ufoevidence.org/documents/doc1447.htm

[xci] Dr. James Deardorff, "A Possible Extraterrestrial Strategy for Earth," in the book Alien Update, edited by Timothy Good

[xcii] The Fermi Paradox is described on the UFO Evidence website: http://www.ufoevidence.org/topics/Fermi.htm

[xciii] Carl Sagan, The Cosmic Connection, Cambridge University Press, 2nd Edition, 2000

[xciv] Brookings report can be found on the Enterprise Mission website: http://www.enterprisemission.com/brooking.html

[xcv] "When UFOs Arrive" Popular Mechanics, February 2004 cover story: http://www.popularmechanics.com/science/space/1283081.html

[xcvi] Clarke's Third Law: see Arthur C Clarke's Laws: http://www.lsi.usp.br/~rbianchi/clarke/ACC.Laws.html

[xcvii] There seems to be conflicting/ambiguous indications as to whether or not the close encounter phenomenon is increasing in scope. In the Roper Poll of Unusual Personal Experiences: 1991 and 1998 http://www.nidsci.org/news/roper_surveys.html, there are indications that there is no such increase. However, David Jacobs in The Threat indicates that there is a systematic increase in the scope of contact. Thus, the numerical dynamics of contact, like most other aspects of this phenomenon, remains ambiguous.

[xcviii] Reference to "The Cosmic Kindergarten" by Stanton Friedman in the site: The UFO Challenge, found on the web at: http://www.v-j-enterprises.com/sfchlng.html

[xcix] The Zoo Hypothesis is defined as the theory that ET visitors are observing and covertly managing humanity. It is described on the website: http://www.physics.hku.hk/~tboyce/sfseti/53sociable.html

[c] One name often used in referring to the visitors is The Watchers. See Raymond Fowler's book The Watchers for a development of this idea.

[ci] Books on alleged implants during UFO abduction experiences, written by Dr. Roger K. Leir:
- Casebook: Alien Implants (Whitley Strieber's Hidden Agendas series)
- The Aliens and the Scalpel : Scientific Proof of Extraterrestrial Implants in Humans

[cii] References to family line relationships among abductees are found in books such as Raymond Fowler's The Watchers, and David Jacobs book Secret Life

[ciii] Budd Hopkins, Witnessed, Pocket Books, 1997
Describes an apparent spectacular display of the UFO presence occurred in New York, in an event which Budd Hopkins describes in his book.

Chapter 9

[civ] The book Making Contact edited by Bill Fawcett, discusses many aspects of potential human – ET contact.

[cv] Dr. James Deardorff sums up an overall speculation on "A Possible Extraterrestrial Strategy for Earth," in his article within the book Alien Update by Timothy Good.

[cvi] Several reports can be found online showing the growing level of resources devoted to military power:
- http://www.globalissues.org/Geopolitics/ArmsTrade/Spending.asp
- http://www.cyberdyaryo.com/features/f2002_1009_05.htm

[cvii] A powerful summary of genocidal rampages in modern times can be found at the website: http://www.genocide.org/

[cviii] An excellent book on current concepts on possible star-flight is "The Starflight Handbook" by Eugene F. Mallove and Gregory L. Matloff (Wiley, 1989). This book shows how starflight might be accomplished with physics as we know them today. An excellent site which covers many advanced space propulsion systems is the Advanced Propulsion Concepts website.

[cix] One excellent paper on Zero-point physics is by Bernard Haisch of the Society of Scientific Exploration: These articles can be found at: http://www.calphysics.org/research.html

[cx] Another location on the internet with excellent articles on physics, potentially leading to starflight is Jack Sarfatti's website: www.stardrive.org

[cxi] The National Space Society (www.nss.org), of which this author is a member, is a large advocate of the human expansion into space, including the idea of a new manifest destiny for humanity in space

[cxii] Barbara Marx Hubbard, Conscious Evolution (see previous reference)

[cxiii] Stuart Kauffman, At Home in the Universe, Oxford University Press, 1996

cxiv Barbara Harris Whitfield, Spiritual Awakening (see previous reference)

cxv Lee Carroll and Jan Tober, The Indigo Children, The New Kids Have Arrived, Light Technology Publishing, 1999. Note: This book describes the Indigo Child phenomenon is defined in the book their accompanying website: http://www.indigochild.com/. Note that this book approaches the topic from a somewhat mystical/metaphysical perspective.

cxvi For further details on the Indigo child phenomenon, see the "Indigo Children" section of the Metagifted.org website: www.metagifted.org/topics/metagifted/indigo

cxvii I have heard varying views stated as to whether the Indigo children are a distinct group from other gifted children, and if so, how this or other such groups would be distinguished. To evaluate an "Indigo Hypothesis," such a distinguishing criterion would need to be established.

cxviii Zachariah Sitchin: See earlier references.

cxix While this is a fascinating hypothesis, I believe that there is presently very little evidence to suggest that humanity actually was directly influenced by ET genetic or cultural manipulation. Thus, I remain both open minded and intrigued by the possibility, while skeptical of the actual proposition.

cxx Graham Hancock, Fingerprints of the Gods, Three Rivers Press, 1996

cxxi Such a study would need to determine the exact characteristics which make one an Indigo child. This would clearly establish a subject and control populations, which we could then compare.

cxxii The primary alternative hypothesis counter to the "Indigo" hypothesis would be that those children, whom we refer to as "Indigo" children, are simply subjected to influences of family members. If the parents are close encounter experiencers, (perhaps with new-age beliefs) then the children might be exposed to such ideas early in life. Thus, the "Indigo" influence might be learned, rather than genetic.

cxxiii The Princeton Engineering Anomalies Research (PEAR) Laboratory: http://www.princeton.edu/~pear/
An excellent description of their work can be found in the book by Robert Jahn and Brenda Dunne, Margins of Reality, Harcourt Brace, 1987

cxxiv The Global Consciousness Project is described on the internet at:
http://noosphere.princeton.edu/

cxxv Several readers, upon reading this section, indicated that they have so far
seen no indication of such an improvement in humanity. Perhaps such an
influence is still to come. Additionally, "improvement" might be open to
interpretation by the viewer. An "improvement" in one person's view might be
seen negatively by another.

Appendix B
cxxvi W.I.B. Beveridge, <u>The Art of Scientific Investigation</u>, Vintage Books, 1957

cxxvii J. Allen Hynek, <u>The UFO Experience</u>, Marlow & Co, New York, 1972,
1998

cxxviii Jacques Vallee, <u>Confrontations</u> , Ballantine Books, 1990

cxxix Two of the best treatments of UFO physical evidence are:
- <u>The UFO Evidence</u> by Richard Hall, 1964, National Investigations
    Committee for Aerial Phenomena
- <u>The UFO Enigma</u> by Peter Sturrock, Aspect, 1999